System Safety and Reliability Analysis
Course Notes

Chapter 1
Introduction

Clifton A. Ericson II
Design Safety Solutions LLC
cliftonericson@verizon.net
540-786-3777

© C. A. Ericson II 2014

Outline

Chapters

1. Introduction
2. Systems Theory
3. System Safety
4. System Safety Process
5. System Safety Topics
6. Reliability
7. Reliability Topics
8. Hazard-Mishap Theory
9. Risk Management
10. Hazard Analysis Techniques
11. Hazard Recognition
12. Hazard Mapping
13. Safety and Reliability Math
14. FMEA
15. FTA
16. Software Safety
17. eTree FTA Program

Appendices

A. References
B. Acronyms
C. Exercises

Chapter 1 - Introduction

Course Focus

- System Safety (SyS)
  - Concept and Process
- Reliability (R)
  - Concept and Process
- Hazard - Mishap Theory
  - Hazard concept
  - Risk Theory
- Hazard Analysis for Safety
  - Concept and Methods
- Failure Mode and Effects Analysis (FMEA) for Reliability
  - Definitions & Construction Methodology
- Fault Tree Analysis (FTA) for Safety and Reliability
  - Definitions, Concepts & Construction Methodology

Key Premises

- Safety and Reliability are emergent properties (qualities) of system design
- Mishaps and failures are dependent upon these properties
- Safety and Reliability can be measured and controlled
- Safety and Reliability are intentionally designed-in
- Safety and Reliability are symbiotic (work together)

- Safety is best achieved via a systems approach – System Safety
- MIL-STD-882E defines the system safety process

Chapter 1 - Introduction

Acronyms

- CF - Causal Factor
- CS - Cut Set
- FT - Fault Tree
- FTA - Fault Tree Analysis
- HA - Hazard Analysis
- HCF – Hazard Causal Factor
- HRI – Hazard Risk Index
- HW - Hardware
- MCS - Min Cut Set
- RAC - Risk Assessment Code
- R - Reliability
- SCF - Safety Critical Function
- SSR - System Safety Requirement
- SW - Software
- SwS - Software Safety
- SyS - System Safety

See Appendix A for a more complete list

Why System Safety and Reliability?

Reality:
- Things fail
- Things wearout
- Humans err
- Designs contain flaws/errors
- Systems utilize hazard sources
- Systems perform hazardous operations

Outcomes:
- System failures (Reliability)
- Accidents (Safety)

Consequences:
- Deaths / injuries
- Loss costs
- System performance
- Cost of ownership
- System unavailability
- Liability
- Sales losses

Mishap Prevention:
- Prevention through design
- Design for safety
- Design for reliability

Mishaps can occur if not intentionally countered via design

Chapter 1 - Introduction

definition >>> Safety

- Dictionary
  - The state of being certain that adverse effects will not be caused by some agent under defined conditions. A safe place.
- MIL-STD-882D
  - Freedom from those conditions that can cause death, injury, occupational illness, damage to or loss of equipment or property, or damage to the environment.
- Intent
  - Freedom from unacceptable mishap risk. [more realistic definition]

definition >>> System Safety

- Dictionary – none
- MIL-STD-882D
  - The application of engineering and management principles, criteria, and techniques to achieve acceptable mishap risk, within the constraints of operational effectiveness and suitability, time, and cost, throughout all phases of the system life cycle.
- Intent
  - A specialized engineering discipline for proactively designing-in safety into a product, through mishap risk management methods for all phases of the system lifecycle.
  - A Design for Safety process based on hazard elimination/mitigation.

Chapter 1 - Introduction

definition >>> Mishap

- Dictionary
    - An unfortunate accident; bad luck.

- MIL-STD-882D
    - An unplanned <u>event</u> or series of events <u>resulting in</u> death, injury, occupational illness, damage to or loss of equipment or property, or damage to the environment.

- Intent
    - An actualized hazard, resulting in an undesired event, or events, with loss outcome; the physical manifestation of a hazard.

Note – In system safety the terms *mishap* and *accident* are synonymous

definition >>> Reliability

- Dictionary
    - The quality of being dependable or reliable.

- MIL-HDBK-338B Electronic Reliability Design Handbook
    - The probability that an item can perform its intended function for a specified interval under stated conditions.

- Intent
    - The property of an item being ready for full use when intended.
    - The act of building reliability into a product.

Chapter 1 - Introduction

Reliability vs. Safety

- Safety and reliability are somewhat independent of each other in terms of requirements and design. Yet, their implementation has an impact on one another.
- Reliability is generally not a precondition for safety and vice-versa.
- In general, better reliability enhances safety and better safety enhances reliability. However, there are exceptions.
- Sometimes there is a conflict between safety and reliability. A safer design sometimes reduces reliability and sometimes a more reliable design reduces safety.
- Reliability has a <u>Direct Safety Impact</u> when the failure of an item directly leads to a mishap.
- Reliability has an <u>Indirect Safety Impact</u> when the failure of an item does not directly leads to a mishap, but is a contributor to a mishap involving several factors.
- Reliability is not Safety and Safety is not Reliability, they are separate emergent properties based on system design.

Safety & Reliability Work Together

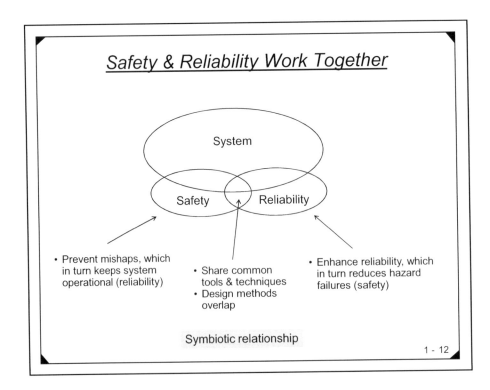

Chapter 1 - Introduction

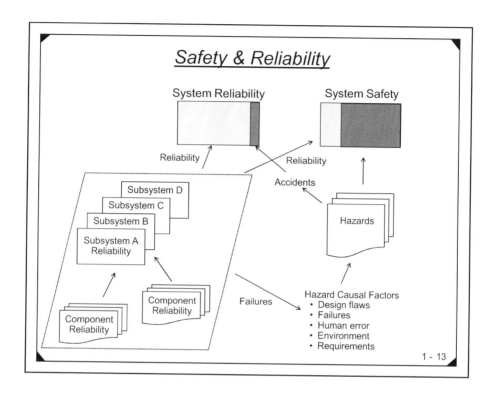

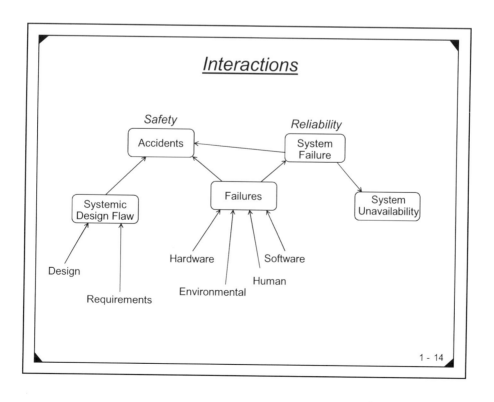

Chapter 1 - Introduction

Hazard vs. Mishap   Safety vs. Reliability

1 - 15

Safety Concepts

- Accidents are predictable and preventable events
  - An accident is not "just one of those random things"
  - Accidents don't "Just" happen, they are pre-determined (through design)
  - Preventing accidents is an option – they are preventable (through design)
- System Safety is the Key
  - System methodology
  - Hazard identification
  - Hazard elimination/mitigation
  - Risk management
- Variables
  - System design
  - System hazards
  - Hazard risk
  - Risk control

1 - 16

Chapter 1 - Introduction

Accidents Don't Just Happen

- An accident is not "just one of those random things"
- They don't have to happen – they are preventable
- Accidents are predictable and preventable events
- Hazards are the key

Accidents are predictable, foreseeable and preventable.
Hazards are the key to predicting potential mishaps.
System safety engineering applies the tools and techniques.

Failures/Errors Are Inevitable

- Hardware will eventually fail
- Human error cannot be prevented

- Some failures & errors are critical, others are not

- Eliminating/mitigating critical failures/error will:
    - Prevent accidents
    - Prevent critical operational failures

The effect of failures/errors can be controlled through Reliability and Safety

Chapter 1 - Introduction

Systemic Faults Are Inadvertently Created

- Systemic failures are system failures that result from design errors rather than actual hardware failures or human error

- Causes include:
  - Design errors
  - Design Flaws
  - Requirements errors
  - Software failures
  - Procedural errors
  - Common cause flaws

The effect of system faults can be controlled through Reliability and Safety

System Safety-Reliability Objectives

- Reliability deals with the task of ensuring that a system performs a required task or mission for a specific time.
  - R = probability that system successfully operates for specified period of time in specified environment

- System safety deals with the task of ensuring that a mishap does not occur within the system while performing its task or mission.
  - S = acceptable hazard risk

- In general, a more reliable system is a safer system and a safer system is a more reliable system.

Chapter 1 - Introduction

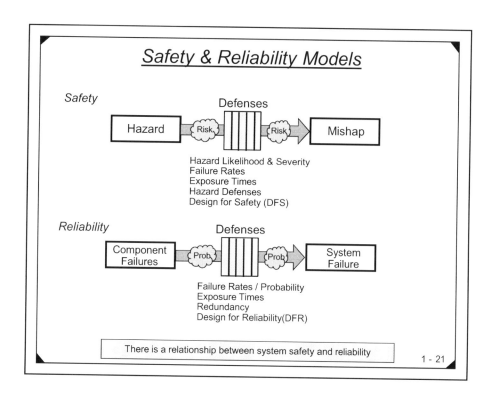

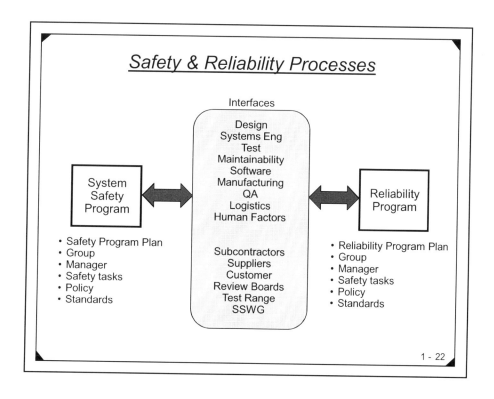

Chapter 1 - Introduction

Chapter 2
Systems Theory

System Safety and Reliability Analysis
Course Notes

Clifton A. Ericson II
Design Safety Solutions LLC
cliftonericson@verizon.net
540-786-3777

© C. A. Ericson II 2014

Why Are Systems Important?

- Life revolves around systems; systems are ubiquitous
- Systems are man-made, as well as Safety and Reliability
- Systems have the potential for mishaps
- Potential mishaps are defined by hazards and hazard risk
- Hazards and risk are the result of system design

- Safety and reliability are created during system development
  - It is necessary to thoroughly understanding system design
  - Must apply S and R during the design process
  - A total systems approach is required

Chapter 2 – Systems Theory

What is a System?

- A collection of inter-related components working together towards some common objective.

- A system may include <u>software</u>, <u>mechanical</u>, <u>electrical and electronic hardware</u> and be <u>operated by people.</u>

- System components are dependent on other system components

- A system is more than simply the sum of its parts!
  - It has a properties of the system as a whole. (emergent properties)

What is a Subsystem?

- A subset of a system.

- A coherent and somewhat independent component of a system.

- A collection of inter-related components working together towards some common objective to support the objectives of the system.

- A subsystem may include <u>software</u>, <u>mechanical</u>, <u>electrical and electronic hardware</u> and be <u>operated by people.</u>

Chapter 2 – Systems Theory

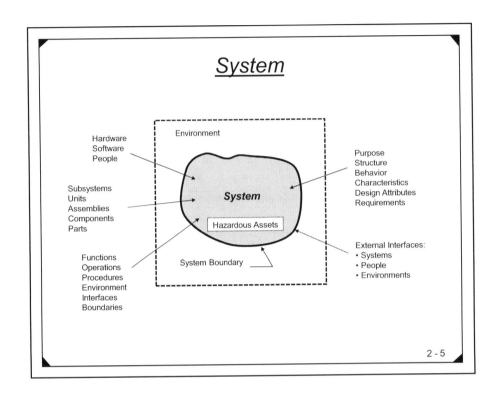

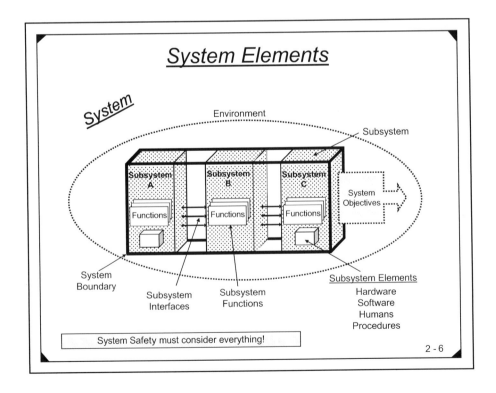

Chapter 2 – Systems Theory

Subsystem Elements

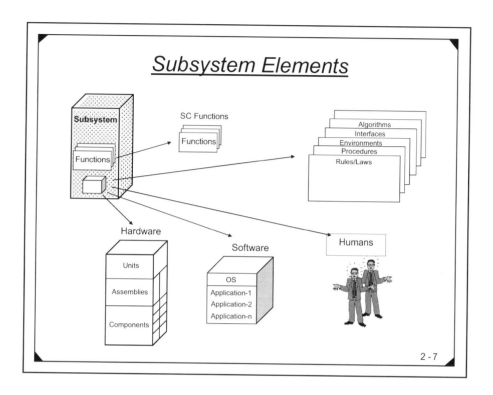

Systems Characteristics

- Structure - defined by components/elements and their composition
- Behavior - involves inputs, processing and outputs of material, energy, information, or data
- Interconnectivity - the various parts of a system have functional as well as structural relationships to each other
- Functions – performs the system tasks
- Rules - governs structure and/or behavior

Chapter 2 – Systems Theory

System Viewpoints

- **Physical** – the architectural view that depicts what the system contains and how it is constructed (ref A).
- **Functional** – describes what the system must do in order to produce the required system behavior, broken down into functions with input, output and transformation rules (ref A).
- **Operational** – defines how the user will interface with, and operate the system, including instructions, conditions, parameters and limitations (ref A).
- **Software** – this view looks at the system software that essentially controls computer controlled systems.
- **Environment** – this view looks at the various environments that the system will encounter (internal and external).
- **Human** – this view looks more closely at human performance in the system and the effect of potential human errors.
- **Organizational** – this view considers the organizational and management aspects affecting a hazard.

Ref A: Alphonse Chapanis, *Human Factors in Systems Engineering*, New York, Wiley, 1996

System Attributes

Elements	Hierarchy	Domains	Operation	Types	Properties
Hardware Software Humans Rules/Laws Environments Interfaces Procedures	Systems Subsystems Units Assemblies Components	Boundaries Limitations Complexity Criticality	Functions Modes Phases Tasks	Static Dynamic Homeostatic Cybernetic	Reliability Quality Maintain. Perform. **Safety**

Systems are complex

System Safety must consider everything!

Chapter 2 – Systems Theory

Example Systems

System	Purpose	Subsystems
Ship	Transport and Deliver Weapons	Engines, Hull, Radar, Communications, Navigation, Software, Fuel, Humans
Aircraft	Transport and Deliver Weapons	Engines, Airframe, Radar, Fuel, Communications, Navigation, Software, Humans
Missile	Deliver Ordnance	Engines, Structure, Radar, Communications, Navigation, Software
Automobile	Transportation	Engine, Frame, Computers, Software, Fuel, Humans
Nuclear Power Plant	Deliver Electrical Power	Structure, Reactor, Computers, Software, Humans, Transmission Lines, Radioactive Material
Television	View Video Media	Structure, Receiver, Display, Electrical Power
Toaster	Browning of Bread	Structure, Timer, Electrical Elements, Electrical Power
Telephone	Communication	Structure, Receiver, Transmitter, Electrical Power, Analog Converter

System Lifecycle Stages

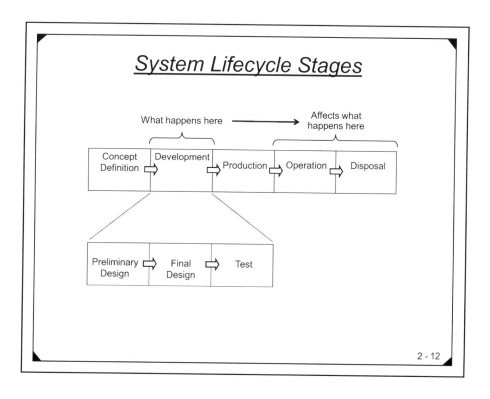

Chapter 2 – Systems Theory

System Development

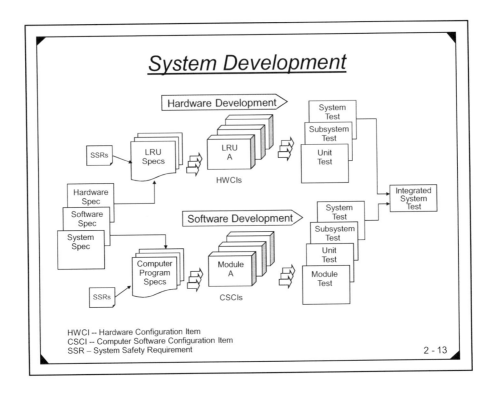

HWCI -- Hardware Configuration Item
CSCI -- Computer Software Configuration Item
SSR -- System Safety Requirement

Systems Engineering (SE)

- SE is the activity of specifying, designing, implementing, deploying, maintaining systems, which include hardware, software, people and interaction of the system with users and its environment
  - SE unites all aspects of engineering to develop a system
  - Processes
  - Tools
  - Timelines

Chapter 2 – Systems Theory

System Safety & SE

- System safety is an element of SE
- System safety is an emergent property of a system
- System safety ensures a system design is safe
- System safety (and hazard analysis) utilizes SE tools
  - Specifications / requirements
  - Simplified system diagrams
  - Functional flow diagrams
  - Reliability block diagrams
  - FMEAs

SE and System Safety are partners in system development

Systems Have Emergent Properties

- These relate to the behaviour of the system in its operational environment
- Properties of the system as a whole rather than properties that can be derived from the properties of system components.
- Emergent properties are a consequence (result) of the relationships between system components
- They can therefore only be assessed and measured once the components have been integrated into a system
- They can change when system configuration changes (note that failures often indirectly change the system configuration)
- Examples:
  - Safety
  - Reliability
  - Quality

Hazard mitigation modifies the safety property

Chapter 2 – Systems Theory

Laws of Systems

- Law #1 - Systems are entities comprised of hardware, software, humans, rules/laws, procedures and environments.
- Law #2 - Systems are designed to work in an intended manner for an intended purpose.
- Law #3 - Systems react predictably to input; they also react predictably to failures (determining these responses are the task of system safety and reliability).
- Law #4 - System size and complexity directly impact system understanding, reliability and safety.
- Law #5 – Systems are designed to perform specific functions when intended during normal operation; however, failure modes can cause these functions to occur inadvertently when not intended.
- Law #6 – Because of component inter-dependencies, faults can be propagated through the system, so failure in one component can affect the operation of other components.
- Law #7 – System failures often occur because of unforeseen inter-relationships between components.
- Law #8 – Systems often have unintended functions accidentally designed into the system.
- Law #9 – If anything can go wrong…it will (Murphy's Law).
- Law #10 – Everything is a system, and every system is part of a larger system.
- Law #11 – Systems have emergent properties (e.g., safety, reliability, quality), which can vary depending upon the system design.

Laws of Systems – Reliability

- Law #1 - A system is reliable when it works as intended for a specified time

- Law #2 - A system is unreliable when it fails to work as specified or intended

- Law #3 - Systems become unreliable due to failures, errors & faults

- Law #4 – System size and complexity drives
  - The number of possible failure components / failure modes
  - The criticality of various failures

- Law #5 - Not all failures result in safety problems or hazards, whereas most failures impact reliability

- Law #6 - When a component fails, the system design (and operation) effectively changes

Chapter 2 – Systems Theory

Laws of Systems – Safety

- Law #1 - Every system contains inherent potential safety vulnerability
- Law #2 - Safety vulnerability is characterized in terms of hazards, mishaps and risk
- Law #3 - Hazards are a unique system attribute with identifiable and controllable components
- Law #4 - Risk is the reference control measure for controlling hazards and potential mishaps
- Law #5 - A system is safe when all mishap risk is known, understood, mitigated and judged as acceptable

Summation of Laws

- Systems have 3 possible major success/fail states
  - Good - system works correctly as intended } Success Space
  - Bad - system failures occur, but in a somewhat safe manner } Failure Space
  - Ugly - system fails badly, everything goes wrong, safety mishaps occur
  - Systems become unsafe due to failures and design flaws
    - Poor design, lack of foresight, interface errors, human errors, sneak paths, etc.
- *Success Space* involves thinking in terms of SUCCESS
  - Design, Performance
- *Failure Space* involves thinking in terms of FAILURE
  - Reliability, Safety
- Safety engineers require a different viewpoint than designers
  - Failure Space vs . Success Space

Chapter 2 – Systems Theory

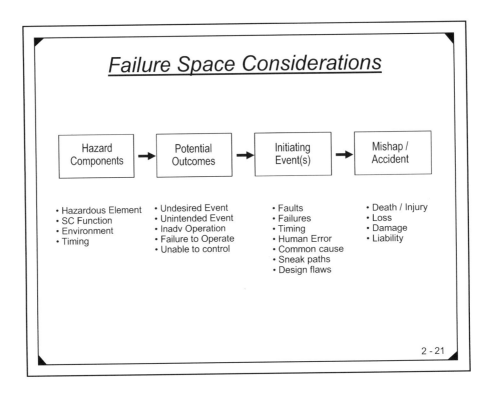

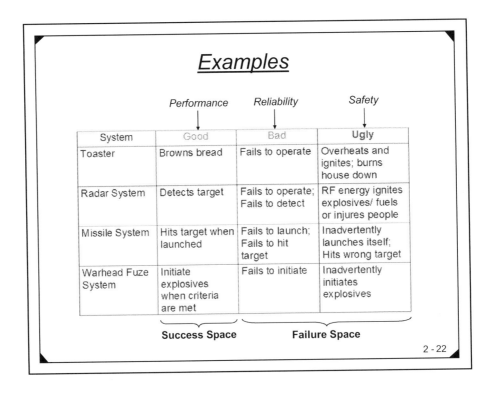

Chapter 2 – Systems Theory

Understanding Failure Space is Important

- System responses are predictable to given input/stimuli
  - Normal data
  - Failures

- Hazards are predictable
  - The key is being able to read the failure space map

- What is predictable (i.e., hazards) can be controlled
  - Mishap risk is reduced by mitigating hazards

- Mishap causes tend not to be just single failures, but a symphony of failures errors, and latent design flaws

System Related Safety Concerns

System Element	Safety Concern
Hardware	Hazardous Elements/Components Hazardous Operations Inadvertent operation Failures Combinations of Faults
Software	Inadvertent Operation Incorrect Control
Humans	Incorrect Operation Incorrect Decisions
Functions	Fail To Perform Malfunction (incorrect) Unintended Function
Interface	No Data Incorrect Data Data Latency
Environment	Severe Weather Hazardous Environments
Procedures	Hazardous Tasks Incorrect Instructions

Chapter 2 – Systems Theory

System Development Artifacts

- System specification and design requirements
- Drawings / schematics
- Simplified System Diagram
- Functional diagrams
  - Function Flow Diagrams
  - Functional Flow Block Diagrams (FFBD)
  - Functional Dependency Diagrams
- System Hierarchy Table or SET
- Reliability Block Diagram (RBD)
- Failure Mode and Effects Analysis (FMEA)
- Timeline Analysis

Functional Flow Diagram

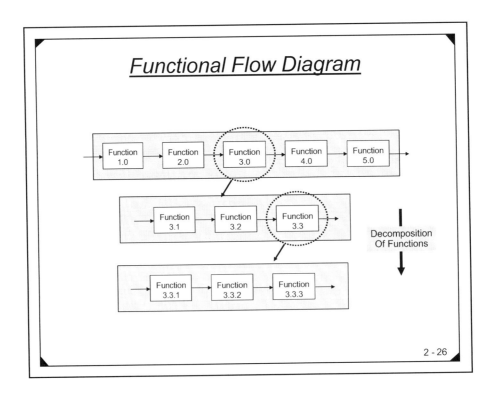

Chapter 2 – Systems Theory

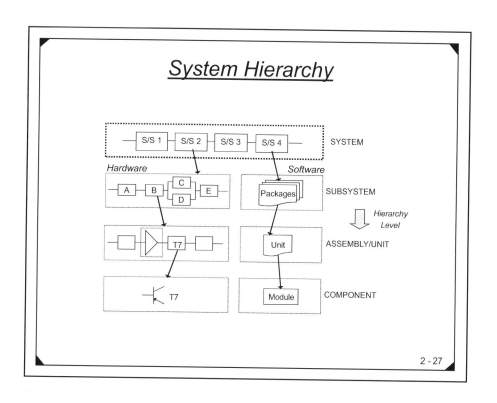

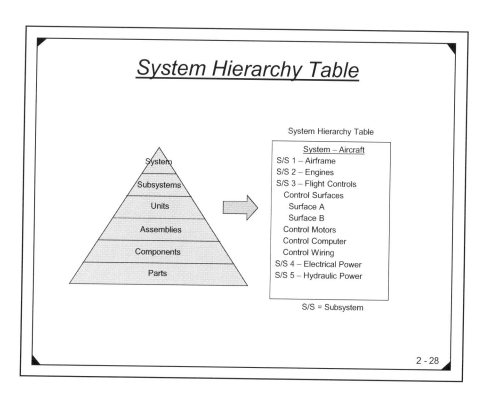

Chapter 2 – Systems Theory

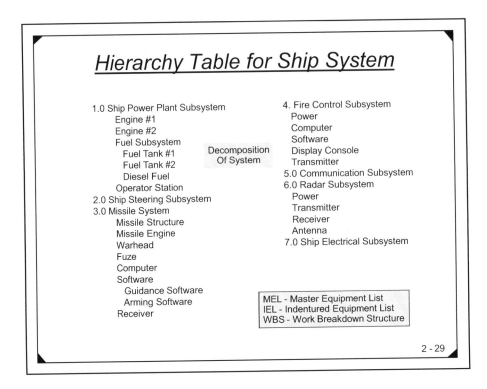

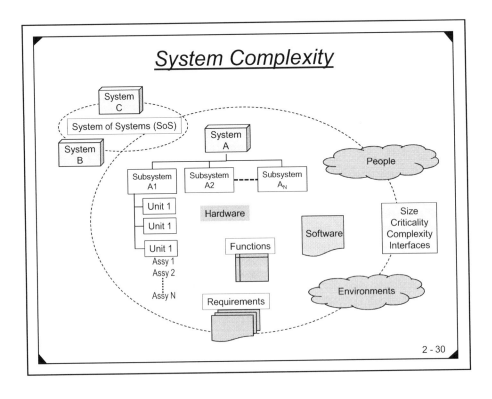

Chapter 2 – Systems Theory

System Complexity

There four basic "system type" models describe almost all types of systems and their relative complexity. Each model varies depending upon factors such as composition, relationships, intent and environment.

These system types are:
a) static
b) dynamic
c) homeostatic
d) cybernetic

The system type typically reflects not only complexity but also safety and reliability.

Source: Systematic Systems Approach, Thomas H. Athey, Prentice-Hall Inc., 1974

2 - 31

System Complexity

Static Systems
Static systems are simple, essentially "dumb" systems that operate only as built or designed. They receive no input information to modify their process and they are unable to modify their process on their own. They have fixed goals, but no means of internal control to ensure that the system goals are met. An example of a static system would be a watch or a clock. The functional purpose is to keep accurate time, but the system has no internal control devices for ensuring this goal is achieved. An ordinary, analog watch has no way to readjust itself if the time is incorrect. Also, the watch has no awareness of changes, such as changing the time to daylight savings time. The environmental effect on the watch may produce different outcomes

Dynamic Systems
Dynamic systems are a little more complex, but are still essentially dumb systems, in that they can only respond as they are programmed to function according to given input information. The system merely provides output according to the input received. An example of a dynamic system would be a computer display. Its goal is to provide a specific output according to the input. It operates according to a fixed set of unchangeable rules. Environmental obstacles, such as water on the electronics, might produce different outcomes

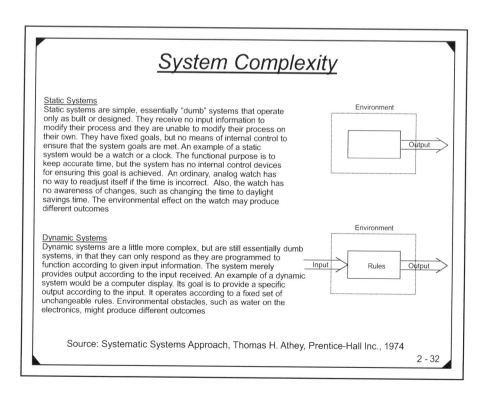

Source: Systematic Systems Approach, Thomas H. Athey, Prentice-Hall Inc., 1974

2 - 32

Chapter 2 – Systems Theory

System Complexity

Homeostatic Systems
Homeostatic systems are a step up in complexity with some level of "intelligence". These systems respond to environmental changes and have effective internal control devices that enable the system to meet its fixed system goals. An example of a homeostatic system would be the internal temperature control system in a house. The house temperature depends upon two main inputs 1) the temperature outside the house, and 2) how much heating or cooling the central system is providing. The output depends on the combined inputs compared to an established standard. If the internal house temperature does not meet the established standard, then the system makes the necessary adjustments until the standard is achieved. This type of system functions to meet the fixed system goals. Another example of this type system would be a guided missile system

Cybernetic Systems
Cybernetic systems are the most complex and intelligent systems. These systems are affected by environmental shifts and have effective internal control devices that enable the system to meet its fixed system goals. In this type of system the goals are not rigidly fixed but are adaptable to changing environments and conditions. An example of a cybernetic system would be a manufacturing process system that has adaptable standards and can adjust or correct both the input and the process itself. These systems tend to typically be safety-critical and require significant system safety and software safety effort

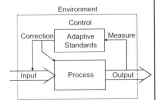

System Complexity

Factor \ Type	Static	Dynamic	Homeostatic	Cybernetic
Complexity	Low	Medium	Medium-High	High
Goals	Fixed	Fixed Rules	Fixed Standards	Adaptive Standards
Output Correction	None	Manual	Self	Self
Safety Criticality	Low	Medium	Medium-High	High
Safety Effort	Low	Medium	Medium-High	High
Reliability Effort	Low	Medium	Medium-High	High

Chapter 2 – Systems Theory

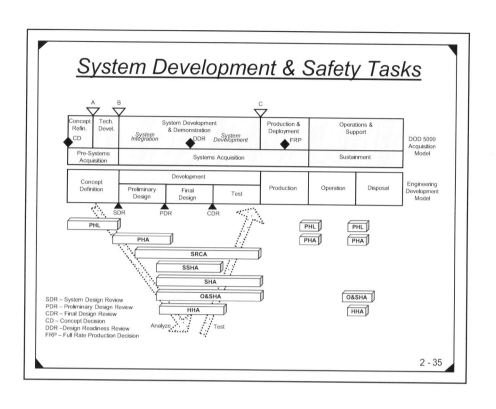

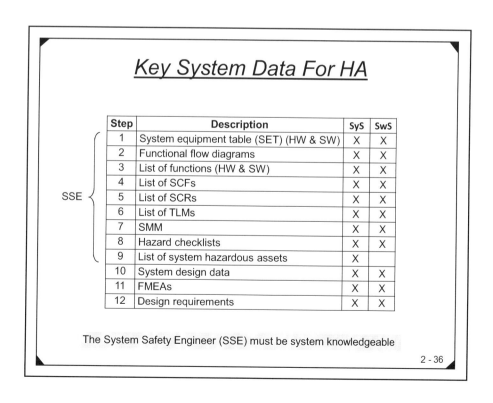

Chapter 2 – Systems Theory

System Safety and Reliability Analysis
Course Notes

Chapter 3
System Safety

Clifton A. Ericson II
Design Safety Solutions LLC
cliftonericson@verizon.net
540-786-3777

© C. A. Ericson II 2014

The Need for System Safety

- Problem
  - Mishaps (accidents) occur
  - Mishaps cost
    - Lives, money, time, litigation, reputation ……..
- Fact
  - Mishaps are a *system* phenomena
  - Hazard and Mishap risk cannot always be eliminated
  - Potential mishaps are predictable and controllable (via hazards)
- Solution
  - Apply a *systems* approach to reduce mishap risk
  - Intentionally build safety into the system from the start
    - Anticipate problems and Design for Safety (DFS)

System safety controls: Hazards >> Risk >> Mishaps

Chapter 3 – System Safety

Mishap Facts

- Mishaps can happen to anyone
  - People within the system
  - People external to the system
- Mishaps can result in varying consequences and severity levels
- Mishaps are created by system designs
  - Every system design produces unique hazards
  - Hazards lead to mishaps
- Mishaps occur as a function of risk
  - Hazards produce mishap risk
  - Actual mishap risk is not always known by users
- Mishap risk can be controlled
  - Hazards (potential mishaps) can be predicted
  - Hazards can be mitigated through special design

Mishaps Can Occur In Any System

- Power Industry
  - Nuclear, coal, hydro, solar, wind
- Transportation Industry
  - Cars, trucks, trains, ships, aircraft, airports, bridges
- Food Industry
- Medical Industry
- Defense Systems
  - Elements – bombs, missiles, guns, directed energy
  - System of systems – net centric (sensors, C^4, delivery systems)
- Farm Industry
- Sports Industry
- Homes

Some systems present greater risk than others

Chapter 3 – System Safety

Well Known Mishaps

- Three Mile Island, Chernobyl - nuclear power
- Titanic - ship
- Bhopal - chemical plant
- Mt. Erebus crash - sight-seeing aircraft
- John Denver crash - private aircraft
- Hindenburg - hydrogen airship
- Therac 25 - medical equipment
- eColi hamburgers - food industry
- Doha, U.S.S. Forestall - military weapon systems
- Apollo, Challenger, Columbia - spacecraft
- Tacoma Narrows Bridge – transportation system
-
-

These mishaps were preventable

Basic Safety Theme

- Hazards will occur in system design (system law)
- Hazards result in mishaps when actualized
- Many hazards in a system cannot be eliminated
- When hazards exist the potential for mishaps also exists
- Mishap potential is defined by mishap risk
- Mishap risk can be controlled via hazard mitigation

Mishap risk is controlled via the System Safety process

Chapter 3 – System Safety

Safety

- Freedom from those conditions that can cause death, injury, occupational illness, damage to or loss of equipment or property, or damage to the environment (MIL-STD-882D).

- Something is safe when:
  - It results in no mishaps, **OR**
  - The risk of a potential mishap is acceptable

Acceptable to who and by what criteria :
- Society or individual
- Government or corporation
- High risk takers vs. Risk adverse

System Safety

The application of engineering and management, principles, criteria, and techniques

to achieve acceptable mishap risk

within the constraints of operational effectiveness and suitability, time, and cost

throughout all phases of the system life cycle.

Source: MIL-STD-882D
The application of engineering and management, principles, criteria, and techniques to achieve acceptable mishap risk within the constraints of operational effectiveness and suitability, time, and cost throughout all phases of the system life cycle.

Chapter 3 – System Safety

System Safety Objective

- Influence design
    - As early as possible in system lifecycle
    - Build safety into the system or product (vice adding later)
    - To eliminate OR mitigate hazards
    - To reduce mishap risk to acceptable level
    - Use best standard practices (MIL-STD-882 or equivalent)

Design for Safety (DFS)

System Safety

- System Safety **IS**:
    - The application of a rigorous defined process
    - Hazard, mishap and risk focused
    - System oriented (considers everything)
    - Proactive
    - A team approach

- System Safety **IS NOT**:
    - A review of accident statistics
    - Reliability
    - Rote compliance with codes/requirements
    - Just looking at failure hardware modes

Chapter 3 – System Safety

System Safety Scope

- System Safety **Includes**:
  - Hazards contributing to death/injury of operating personnel, system loss, equipment damage, and environmental damage
  - Consideration is given to:
    - Normal and abnormal system operation
    - Normal and abnormal environments
    - Non-operational persons in vicinity

- System Safety **Does Not Include**:
  - Hazards contributing to loss of system operational effectiveness
  - Hazards contributing to loss of system self defense

What IS a System Safety Issue?

- Ship Example:
  - Personnel
    - Technician electrocuted while conducting maintenance
    - Personnel asphyxiated due to insufficient ventilation from rocket launch exhaust gases
  - Own-Ship
    - Equipment damage due to high temperature condition
    - Missile launch into own-ship superstructure
    - EMR from radar ignites aircraft fuel
  - Friendly Asset
    - Friendly aircraft identified as hostile
    - Tactical missile launched during training exercises
  - Major Mishap
    - Inadvertent missile launch

Chapter 3 – System Safety

What IS NOT a System Safety Issue?

- Examples:
  - Operational Effectiveness
    - Hostile target identified too late
    - System fails to operate due to failures
  - Survivability
    - System failure prevents engagement of enemy
    - Damage to own system or personnel by enemy weapon systems

Importance of System Safety

- <u>Saves</u> lives and dollars

- <u>Prevents</u> or <u>controls</u> mishaps

- Methodology for <u>risk management</u>
  - Understand, mitigate, communicate, accept

- Methodology for <u>designing safety into</u> a product
  - Proactive – Foresight vs. hindsight

- Benefits
  - Customer, user, company, society

System Safety is consistent with Integrity and Ethics

Chapter 3 – System Safety

Approaches To Safety

- No Safety
  - Avoid; Do nothing; Protect with insurance
  - Gamble on future mishap probabilities and loss costs
- Program Safety
  - Laws; regulations
  - Slogans' signs
- Safety After the Fact – Reactive process
  - Safety is avoided until needed
  - Safety is added following litigation
  - Safety is added following a mishap (Fly-Fix-Fly concept)
- Safety Before the Fact – Proactive process
  - Safety is intentionally designed into product (to avoid mishaps)
  - Investment in the future
  - Anticipate and mitigate mishap risk
  - System Safety process

Before System Safety

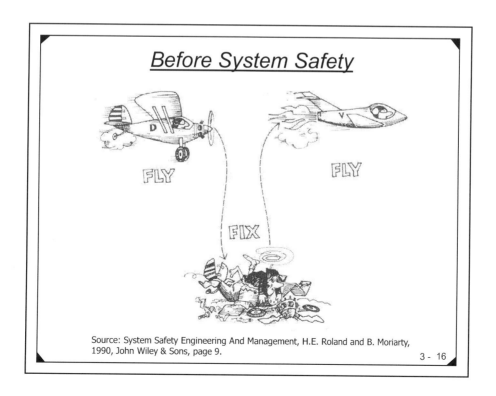

Source: System Safety Engineering And Management, H.E. Roland and B. Moriarty, 1990, John Wiley & Sons, page 9.

Chapter 3 – System Safety

System Safety Considerations

- Prevent mishaps (proactively)
- Reduce hazard/mishap risk to personnel and equipment
- Improves operational availability and system readiness
- Reduces operating and support costs
- Long term investment (save mishap costs)
- Reduces product liability

Struggle

⟵─────────────────────────⟶

Critical Thinking	vs.	Human Nature
• Common Sense		• Funding (Cost & Profit rule)
• Good business		• Schedule
• High Integrity		• Low Integrity
• Ethics		• Lack of Knowledge

3 - 17

Implementing System Safety

- Government
  - Law
  - Directives
  - Guidelines

- Business
  - Require by law
  - Required by contract
  - Common sense (Integrity & Ethics)
  - Economic sense

> Typically there is resistance to system safety unless it is a requirement, or there has been a mishap.

3 - 18

Chapter 3 – System Safety

Standards Are Not Enough

- Safety standards are beneficial and necessary
- But, they only provide a minimum baseline level of safety
- Something more is needed for full safety coverage
- Example: Standard requires triple redundant hydraulic lines for aircraft flight control system
  - This provides high reliability and safety
  - However, SPF can wipe out triple redundancy (e.g., DC-10)
  - A proper Hazard Analysis would have detected this design flaw
- A <u>System Safety Program</u> is needed, over and above safety standards in order to ferret out the hidden hazards unintentionally created

Where Standards Have Failed

- Titanic – individual sealed compartments
  - Sort of .....
- DC-10 – triple redundant hydraulic lines
  - Except for .....
- Lineman extension ladder/buckets – watch for power lines
  - Human error; failure to remember rules
- Carnival cruise ship – redundant engines
  - Four engines in one compartment
-
-
-

Chapter 3 – System Safety

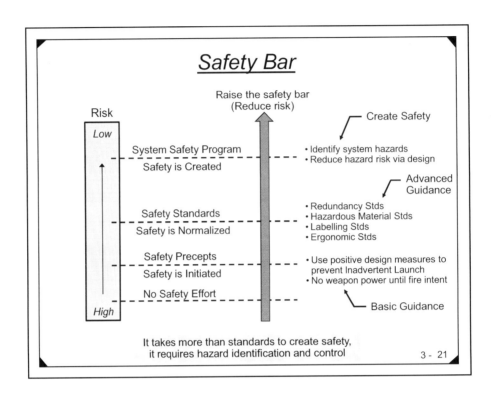

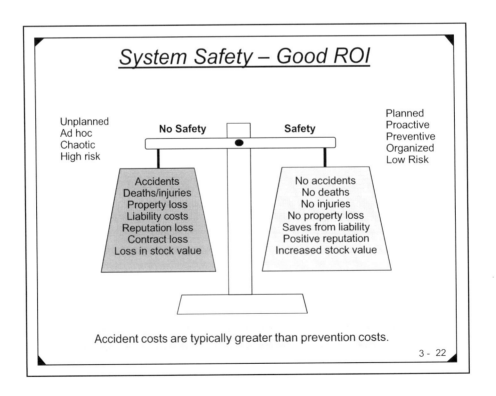

Chapter 3 – System Safety

Galloping Gertie

3 - 23

Fill Her Up Please!

3 - 24

Chapter 3 – System Safety

System Safety - History

- System safety has been an <u>established process</u> since the early 1960's
  - USAF Ballistic Missiles Division – Minuteman Weapon System
  - MIL-S-38130 and MIL-STD-882
- The techniques and requirements have evolved to meet industry needs
- The process has been recognized throughout history
  - Code of Hammurabi (Babylon) – If the house falls in killing the occupants, the builder shall be put to death
  - An ounce of prevention is worth a pound of cure
  - Safety must be designed and built into airplanes, just as are performance, stability, and structural integrity". [Stieglitz, 1948]
  - The safety group must be just as important a part of the organization as stress, aerodynamics, or weight groups, and must have as much influence on all design decisions". [Stieglitz, 1948]

System Safety - History

- SyS began as a formal engineering discipline circa 1960
- Primarily started as a result of military demands for safety of high consequence weapon systems and aircraft
- MIL-STD-882 developed the <u>standard practice</u>
- Approach has been adopted by many other industries
  - Commercial aircraft (SAE ARP-4754 and SAE ARP-4761)
  - Rail transportation
  - Nuclear power
  - Automobile
  -
  -

A proven and successful process

Chapter 3 – System Safety

MIL-STD-882 Improvements

- MIL-S-38130   6 June 1966   9 pages
- MIL-STD-882   15 July 1969   24 pages
- MIL-STD-882A 28 June 1977   32 pages
- MIL-STD-882B 30 March 1984   99 pages
- MIL-STD-882C 19 January 1993   121 pages
- MIL-STD-882D 10 February 2000   31 pages
- MIL-STD-882E (11 May 2012)   104 pages

Version C plus

Result of acquisition reform

MIL-STD-882 is the best practices standard for System Safety

3 - 27

System Safety Scope

- System Safety **IS**:
    - The application of a rigorous defined process
    - Hazard, mishap and risk focused
    - System oriented (considers everything)
    - Proactive
    - A team approach

- System Safety **IS NOT**:
    - Reliability or quality
    - A review of accident statistics
    - Rote compliance with codes / standards
    - Just looking at hardware failure modes

3 - 28

Chapter 3 – System Safety

System Safety Scope

- System Safety **Includes**:
  - Hazards contributing to death/injury of operating personnel, system loss, equipment damage, and environmental damage
  - Consideration is given to:
    - Normal and abnormal system operation
    - Normal and abnormal environments
    - Non-operational persons in vicinity
  - Counters failures, human errors and requirement errors

- System Safety **Does Not Include**:
  - Hazards contributing to loss of system operational effectiveness
  - Hazards contributing to loss of system self defense

System Safety Objectives

1) Explicitly define safety and system safety as a <u>core value</u>
   - The company and organization must stand behind this value
   - Establishes a system safety culture
   - Use a technically qualified system safety staff
   - Management must ensure there is proper funding and resources for the SSP

2) Prevent any <u>initial</u> unnecessary hazards in the system design
   - Identify hazards and design-out those hazards that can be effectively removed
   - Sensitize design engineers to be attentive to system hazards while creating the design, so they may minimize the number of hazards initially residing in the system

3) Manage <u>residual</u> hazards (risk) that cannot be eliminated
   - Residual hazards always exist in the system design for valid reasons
   - This is achieved by mitigating the residual risk of hazards that cannot be eliminated and assuring the proper level of management awareness and acceptance of these risks

Chapter 3 – System Safety

Safety Layers of Defense

The basic system safety philosophy for achieving this goal is to confront hazards, mishaps and risk at three different levels of safety defense:

- Design the system to operate safely under <u>normal</u> operating conditions.
- Design the system to safely tolerate <u>abnormal</u> operations caused by faults and errors.
- Design the system to provide <u>survival protection</u> from foreseeable mishaps.

The System Safety Process

- Document system safety <u>plan for the program</u>
- Identify <u>hazards</u>
- Assess mishap <u>risk</u>
- Apply risk <u>mitigation</u> measures (reduce risk to an acceptable level)
- Reduce risk by selecting best mitigation alternatives
- Verify mishap risk <u>mitigation</u> (test evidence)
- Accept residual mishap <u>risk</u> (transparency)
- Track <u>hazards</u>, data, mitigations, closures and risk

A best practices approach per MIL-STD-882

Chapter 3 – System Safety

System Safety Breakdown (882E)

- System Safety - The application of engineering and management principles, criteria, and techniques to achieve acceptable risk within the constraints of operational effectiveness and suitability, time, and cost throughout all phases of the system life-cycle.

  ↙— Technical aspects (haz identification, risk analysis, mitigation, etc.)

- System Safety Engineering - System safety engineering. An engineering discipline that employs specialized knowledge and skills in applying scientific and engineering principles, criteria, and techniques to identify hazards and then to eliminate the hazards or reduce the associated risks when the hazards cannot be eliminated.

  ↙— Management aspects (planning, staffing, schedule, etc.)

- System Safety Management - All plans and actions taken to identify hazards; assess and mitigate associated risks; and track, control, accept, and document risks encountered in the design, development, test, acquisition, use, and disposal of systems, subsystems, equipment, and infrastructure.

Systems Engineering (882E)

- Systems Engineering - The overarching process that a program team applies to transition from a stated capability to an operationally effective and suitable system. Systems Engineering involves the application of SE processes across the acquisition life-cycle (adapted to every phase) and is intended to be the integrating mechanism for balanced solutions addressing capability needs, design considerations, and constraints. SE also addresses limitations imposed by technology, budget, and schedule. SE processes are applied early in material solution analysis and continuously throughout the total life-cycle.

  ↗— System safety is part of, and works with, systems engineering

Chapter 3 – System Safety

System Safety Concept Validation

- Examples of <u>success</u> using system safety
  - Minuteman missile system
  - Morgantown PRT people mover
  - B-1 Terrain Following Radar System
  - Navy UCAS X-47 Unmanned Aircraft

- Examples of <u>inadequate</u> system safety
  - Ford Pinto (1970's)
  - Therac-25 (1985-1987)
  - DC-10 1989 UAL 232
  - Predator UAV Nogales, Arizona (2006)

System Safety Axioms

- System safety is based upon several key axioms
- These axioms establish the concept
- These axioms define the process and methods

Chapter 3 – System Safety

System Safety Axiom #1

An accident cannot occur unless a hazard exists.

- What causes an accident?
- Accidents don't just happen
- An accident is not "just one of those random things"
- They don't have to happen ... they are preventable
- Accidents are predictable and preventable

- Hazards cause accidents

Accidents have specific causal factors.

System Safety Axiom #2

There is a direct link between hazards and mishaps.

- Hazards lead to accidents
- Hazards are the key to predicting potential accidents/mishaps
- The likelihood of an accident depends upon hazard probability
- Accident severity depends upon hazard severity
- Accident risk is based upon hazard risk
- Hazard risk is controlled by design mitigation of hazards

Understanding hazards and mishaps is important for safety.

Chapter 3 – System Safety

System Safety Axiom #3

> Most system hazards <u>must</u> exist; they cannot be eliminated.

- Many systems must utilize hazardous assets

- Examples
    - Fuel in aircraft, ships, autos
    - Explosives in weapons
    - Electricity in power lines, houses, factories
    - EMR produced by electronics
    -
    -
    -

Hazards and risk will always exist.

System Safety Axiom #4

> Most hazards are man-made.

- We naturally create hazards due to the components used
- We inadvertently design hazards into a system
    - Design errors / flaws
    - Interface errors
    - Not countering failures and human errors

- Although we create hazards, they can be controlled
- We have the option to control hazards
- System safety has the processes to control hazards

We design hazards as well as systems.

Chapter 3 – System Safety

System Safety Axiom #5

Accidents can be prevented by controlling hazards and risk.

- Accidents are predictable, foreseeable and preventable
- Hazards are the key to predicting potential accidents/mishaps
- Hazard and risk mitigation prevents mishaps

- Hazard risk can be controlled

Allowing accidents to occur is an option.

System Safety Axiom #6

A hazard is comprised of 3 necessary components.

- Hazard Source (HS)
    - Hazardous element in the system
    - The basic hazardous resource creating the hazard, such as a hazardous energy source like explosives
    - This component is usually hardware, but can be a function, environment, etc.
- Initiating Mechanism (IM)
    - Hazard trigger mechanism(s)
    - This is the mechanism(s) that causes actualization of the hazard from a dormant state to an active mishap, such as a failure mode
    - This component can be HW, SW, human error, environment, procedures
- Target / Threat Outcome (TTO)
    - Mishap outcome and severity
    - The person or thing that is vulnerable to injury and/or damage, such as aircraft loss due to crash
    - This component can be people, systems, environment, animals

All 3 are required for a hazard to exist; the DNA of a hazard.

Chapter 3 – System Safety

System Safety Axiom #7

> Risk is a safety metric for measuring hazards.

- Accidents are predictable, foreseeable and preventable
- Hazards are the key to predicting potential accidents/mishaps
- Hazard and risk mitigation prevents mishaps

- Hazard risk can be controlled

Accidents can be prevented by controlling hazards-risk.

System Safety Axiom #8

> Hazard analysis (HA) is the key to system safety.

- HA is a methodology for the discovery of hazards
  - Rigorous and Systematic
  - Proactive – design out hazards early in development phase
  - Thorough – covers hardware, software and people
  - Provides an integrated systems approach

- HA provides a mechanism for risk control
  - Assess risk presented by hazards
  - Determine if risk must be modified
  - Measures risk mitigation methods

- Many HA techniques are available

Hazards and risk are identified via HA.

Chapter 3 – System Safety

System Safety Axiom #9

Human error mishaps are typically really design errors.

- Humans will naturally err, but most mishaps attributed to human error are actually design problems
  - Dr. Alphones Chapanis, paper titled "To Error Is Human, To Forgive, Design", presented at the 25th ASSE Conference, 1986
- The system must be designed to account for, and counter, potential human errors that are critical to safety

Many designs lead humans to error; human factors is important.

System Safety Axiom #10

Safety is no accident, it must be created.

- Safety is a system quality (feature)
- Safety is never originally inherent in a system…it must be designed-in
- Safety risk is application and design dependent
- System Safety involves a proven process
- System safety is focused on system, lifecycle, hazards, risk
- Residual risk will always exist (from hazards that cannot be eliminated)
- Risk must be identified and measured
- System safety applies to any type of system/product

System safety is a "save-it-forward" concept.

Chapter 3 – System Safety

System Safety Axiom #11

> System safety is a proven process for risk reduction.

- HA is a methodology for the discovery of hazards
  - Rigorous and Systematic
  - Proactive – design out hazards early in development phase
  - Thorough – covers hardware, software and people
  - Provides an integrated systems approach
- HA provides a mechanism for risk control
  - Assess risk presented by hazards
  - Determine if risk must be modified
  - Measures risk mitigation methods
- Many HA techniques are available

> Hazards and risk are identified via HA.

System Safety

- **Safety Risk Mitigation Methods (Bad to Good)**   Bad
  - Ignore
  - Hope & Pray
  - High Level Inspections
  - Buy Insurance - Sell Risk to Someone Else
  - Fly-Crash-Fix-Fly-Crash-Fix…. (Test Safety Into System)
  - Minimal standards
  - **Practice System Safety**
    - Establish a System Safety Program
    - Identify Hazards   Good
    - Conduct Analyses
    - Conduct Tests
    - Mitigate Hazards
    - Determine Residual System Risk
    - Communicate Residual System Risk

Chapter 3 – System Safety

System Safety Overview

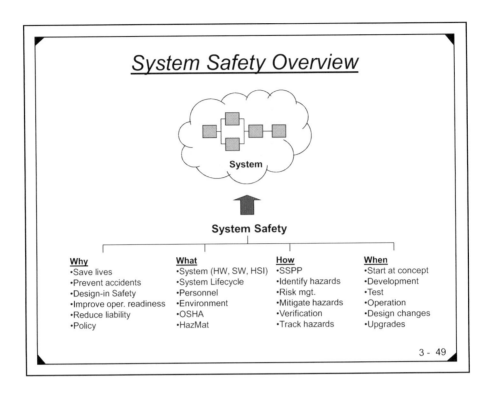

System Safety

Why
- Save lives
- Prevent accidents
- Design-in Safety
- Improve oper. readiness
- Reduce liability
- Policy

What
- System (HW, SW, HSI)
- System Lifecycle
- Personnel
- Environment
- OSHA
- HazMat

How
- SSPP
- Identify hazards
- Risk mgt.
- Mitigate hazards
- Verification
- Track hazards

When
- Start at concept
- Development
- Test
- Operation
- Design changes
- Upgrades

3 - 49

System Safety – Stages of Growth

Stage 1 – Lack of understanding (anger, grief, denial and rationalization)

- Safety is too expensive, there must be a simple cheap way
- Our designers know how to build safe systems
- Statistical mishap data say we are okay
- Surely guidelines and standards are sufficient
- Prevent or fix later – what is the goal?

Stage 2 – Understanding (acceptance)

- Perhaps something proactive and positive is needed
- Safety level unknown (you don't know what you don't know)
- There could be unknown hazards in the system
- Safety can only be characterized via hazards and risk
- The system safety process designs-in safety
- Mishaps can be prevented via hazard mitigation
- Risk can be controlled by design safety features

3 - 50

Chapter 3 – System Safety

System Safety - Stages of Growth

Stage 3 – Evaluate hazard risk and mitigate hazards (implementation)

- Apply the system safety process
- Identify hazards and risk
- Each hazard has different risk (box color and size)
- Some hazards have dependencies (connecting lines)
- High risk hazards are mitigated via design features
- Establish design system safety requirements (SSRs)

Stage 4 – Apply the system safety and HA results (experience)

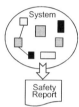

- Apply safety features to mitigate unacceptable risk
- Close each hazard based on V&V evidence
- Only accept risk after successful V&V of SSRs
- Prepare safety assessment report (safety case) summarizing hazard and risk exposure
- Safe system design – all hazards are exposed, evaluated, mitigated, communicated and officially accepted

System Safety - Stages of Growth

Stage 5 – Continuous improvement (company formalization)

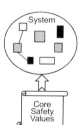

- Make safety a core value
- Develop corporate safety culture
- Develop corporate safety policies, standards, guidelines
- Refine and tailor corporate safety processes
- Develop corporate safety tools
- Conduct corporate safety audits
- Develop corporate safety training

Chapter 3 – System Safety

Basic System Safety Tenets

- System safety is *system* oriented
- System safety deals with "failure space" vs. "success space"
- Reliability, QA, SE, standards are not sufficient for safety
- The amount of System safety effort is application dependent
  - System size, complexity and criticality

- Safety is measured by <u>potential mishap risk</u>
- Hazards create risk
- Hazards can be foreseen and eliminated / controlled
- Risk is mitigated via hazard mitigation (via design safety)

System Safety – Summary

- We live in a fast moving, complex, accident prone world
- Mishaps seem to increase with technology increases
  - Titanic, commercial aircraft, TMI, Chernobyl
- Mishaps are costly (lives, time, money, litigation)
- Mishaps can impact more than those directly involved
  - Bhopal chemical explosion, Chernobyl radiation release
- Potential mishaps are created (unintentional design)
- Mishaps can be controlled via System Safety
  - Systematic approach to manage risk
  - Proactive
  - Cost effective
- MIL-STD-882 is the "standard" best practice for system safety
  - Hazard control via risk mitigation
  - Design-in safety on a timely basis

Chapter 3 – System Safety

Mishaps Are Predictable & Preventable

Babyblues.com - April 13, 2013

Mishaps are:
- not random events
- predictable
- preventable

Hazards are the key to recognition and prevention

3 - 55

System Safety

- System Safety has many aspects and purposes, however, overarching goals of system safety include:
    - To *proactively* identify and eliminate/mitigate hazards
    - To save lives and preclude monetary losses by preventing system mishaps
    - To protect the system and its users, the public and the environment from mishaps
    - To design and develop systems presenting minimal mishap risk
    - To intentionally design safety into the overall system fabric
    - To reduce costs by building-in safety from the start, rather than adding it later (at a higher cost)
- System Safety is effectively a Design-for-Safety (DFS) process, involving technology, discipline and culture
- There is a standardized process

3 - 56

Chapter 3 – System Safety

System Safety and Reliability Analysis
Course Notes

Chapter 4
System Safety Process

Clifton A. Ericson II
Design Safety Solutions LLC
cliftonericson@verizon.net
540-786-3777

© C. A. Ericson II 2014

Background

- System safety is an engineering discipline for intentionally designing safety into a system from the start of development

- System safety has been in practice for 50+ years with a successful track record

- A key standard has been in place for 45 years

- System safety involves 8 key core tasks

- The process can be applied to any system

Chapter 4 – System Safety Process

System Safety

- Management tasks
  - Plan
  - Organize
  - Staff
  - Lead/direct
  - Monitor/Control
  - Motivate
  - Safety culture
  - Policy

- Engineering tasks
  - Hazard identification
  - Risk assessment
  - Hazard mitigation
  - Hazard tracking
  - Design support
  - Develop safety evidence

System safety is a planned proactive process.
System safety is a form of forensic engineering, before the fact.

Problem

System Design → Leads To → Hazards → Lead To → Risk → Leads To → Mishaps

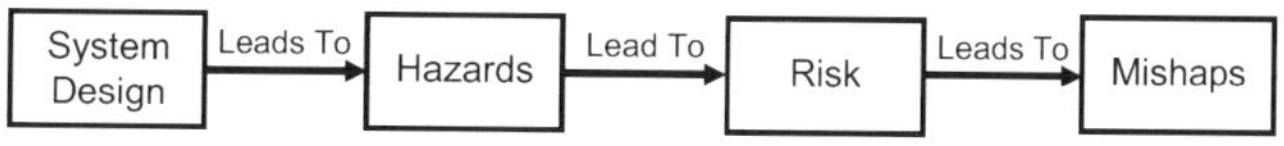

The system design is responsible for mishaps

Chapter 4 – System Safety Process

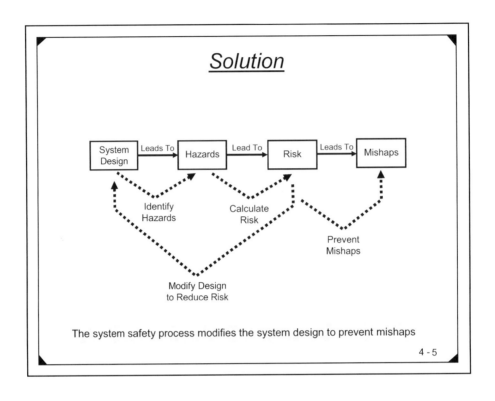

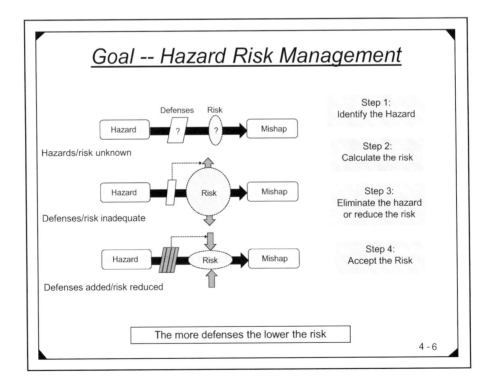

Chapter 4 – System Safety Process

General Requirements (MIL-STD-882)

	882D	882E
1	Documentation of the system safety approach	Document the system safety approach
2	Identification of hazards	Identify and document hazards
3	Assessment of mishap risk	Assess and document risk
4	Identification of mishap risk mitigation measures	Identify and document risk mitigation measures
5	Reduction of mishap risk to an acceptable level	Reduce risk
6	Verification of mishap risk reduction	Verify, validate and document risk reduction
7	Review of hazards and acceptance of residual mishap risk by the appropriate authority	Accept risk and document
8	Tracking of hazards, their closures, and residual mishap risk	Manage life-cycle risk

System Safety Process – Core Tasks View

- Document Safety Plan
- Identify Hazards
- Assess Hazard Risk
- Mitigate Risk
- Reduce Risk
- Verify Risk Mitigation
- Accept Risk
- Track Hazards/Risk

1. Thoughtfully plan and document the safety program approach; obtain program buy-in.
2. Identify hazards and causal factors using a systematic approach.
3. Assess severity and probability of hazard mishap risk presented by hazards.
4. Establish hazard mitigation measures (and requirements); reduce hazard risk to an acceptable level.
5. Reduce hazard risk to an acceptable level by selecting best mitigation alternatives.
6. Verify hazard risk reduction, primarily via requirements verification and testing.
7. Verify and accept hazard risk by the appropriate authorities; close hazards following final acceptance.
8. Record and track hazards, their closures, and residual mishap risk.

MIL-STD-882D/E

Chapter 4 – System Safety Process

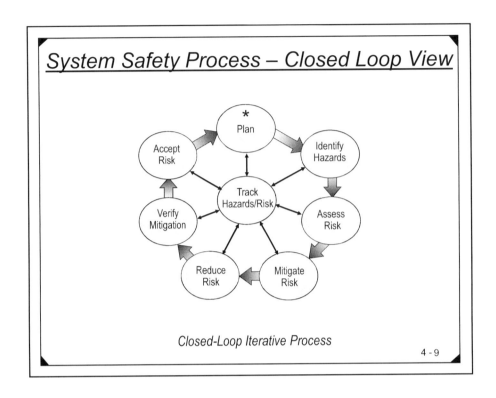

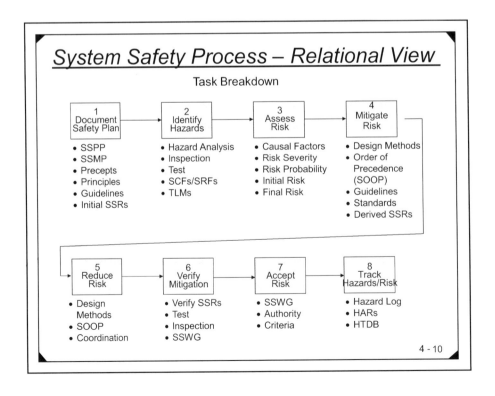

Chapter 4 – System Safety Process

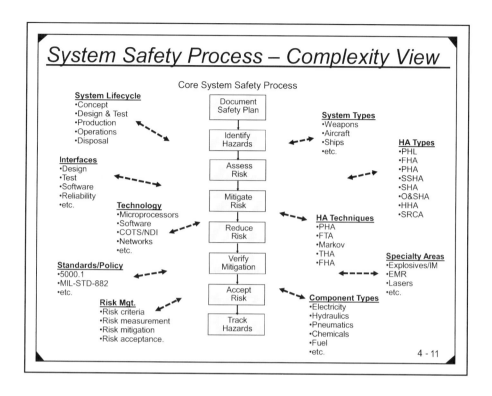

1 - Document Safety Plan

- Plan the overall system safety program (SSP)
- Plan contains tasks, schedules, milestones, products, etc.
- Referred to as a System Safety Program Plan (SSPP)
- Plan can be tailored (from MIL-STD-882)
- Should be approved by management and customer
- Include initial SSRs

Chapter 4 – System Safety Process

2 – Identify Hazards

- Perform hazard analysis (HA) to identify hazards
- Several different HA techniques may be required
- Hazards are documented for traceability and iterative processing
- HA requires skilled knowledgeable people
    - Understand system design
    - Understand what a hazard is
    - Understand how to identify hazards

3 - Assess Hazard Risk

- Risk is a metric used to characterize the criticality of a hazard
- During the HA process the initial risk presented by a hazard is determined
- If the risk is unacceptable, then hazard mitigation is applied to reduce the risk
- Following hazard risk mitigation the final risk is determined
- Hazards are accepted (or not) based on their residual risk

Chapter 4 – System Safety Process

4 – Mitigate Risk

- Identify design mitigation methods to reduce the risk
- There are many types of design mitigation techniques, such as:
  - Redundancy
  - Interlocks
  - Fault tolerant design
- Mitigation methods are applied according to the safety order of precedence (SOOP)
  - SOOP focuses on positive design solutions over less positive human solutions, such as training
- Design safety mitigations are written into design requirements for tracking, testing and validation
  - Design safety requirements (DSRs)
  - System safety requirements (SSRs)

5 – Reduce Risk

- If the risk presented by a hazard is unacceptable, then design safety methods are applied to reduce the risk
- Utilize the risk tables to determine risk level
- Mitigation methods are applied according to the safety order of precedence (SOOP)
  - SOOP focuses on positive design solutions over less positive human solutions, such as training

Chapter 4 – System Safety Process

6 – Verify Risk Mitigation

- Design safety mitigations exist as SSRs for design implementation
- SSRs are tracked to verify implementation
- SSRs are tested to validate implementation and successful use
- When SSRs are successfully validated the final hazard risk can be documented

7 – Accept Risk

- System risk should always be known and understood
- System risk is determined via hazard risk
- The risk of every hazard should be known and accepted by the appropriate acceptance authorities
    - Management
    - Customer
    - Government
- The higher the risk the higher the acceptance authority
- The risk rating, risk acceptance and risk authority should be documented in the SSPP
- The specific acceptance process is company dependent

Chapter 4 – System Safety Process

8 – Track Hazards / Risk

- Hazard tracking involves collecting all information regarding a hazard and formally documenting this information
- Hazard tracking information is placed into a Hazard Tracking System (HTS)
- Hazard information is used for current program use and historical archiving
- An electronic HTS database is the most effective method for hazard tracking

Support Tasks

- Safety Assessment Report (SAR)
- Documentation
- Design Reviews
- Special Tests
- System Safety Working Group (SSWG)
- Audit support
- Support for review agencies (WSESRB, SSSTRP, LSRB, etc.)
- Technical meetings
- Training

Chapter 4 – System Safety Process

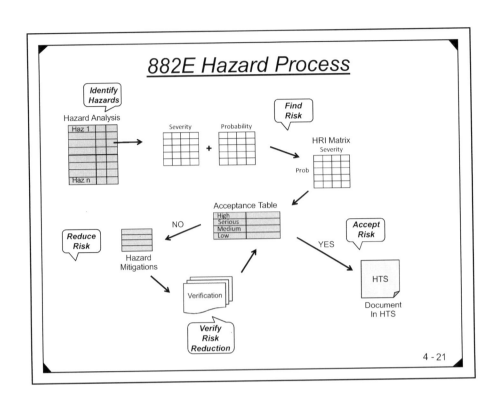

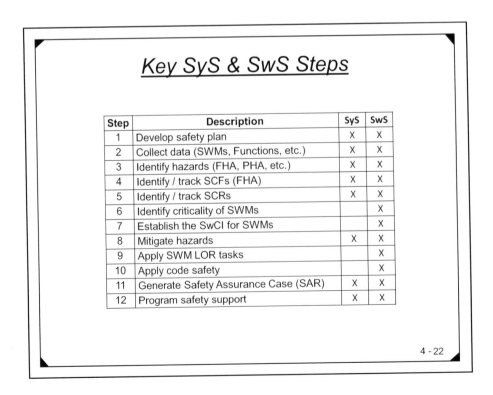

Chapter 4 – System Safety Process

Task Application Matrix 882E

Task	Title	Task Type	MSA	TD	EMD	P&D	O&S
101	Hazard Identification and Mitigation Effort Using The System Safety Methodology	MGT	G	G	G	G	G
102	System Safety Program Plan	MGT	G	G	G	G	G
103	Hazard Management Plan	MGT	G	G	G	G	G
104	Support Government Reviews/Audits	MGT	G	G	G	G	G
105	Integrated Product Team/Working Group Support	MGT	G	G	G	G	G
106	Hazard Tracking System	MGT	S	G	G	G	G
107	Hazard Management Progress Report	MGT	G	G	G	G	G
108	Hazardous Materials Management Plan	MGT	S	G	G	G	G
201	Preliminary Hazard List	ENG	G	S	S	GC	GC
202	Preliminary Hazard Analysis	ENG	S	G	S	GC	GC
203	System Requirements Hazard Analysis	ENG	G	G	G	GC	GC
204	Subsystem Hazard Analysis	ENG	N/A	G	GC	GC	GC
205	System Hazard Analysis	ENG	N/A	G	G	GC	GC
206	Operating and Support Hazard Analysis	ENG	S	G	G	G	S
207	Health Hazard Analysis	ENG	S	G	G	GC	GC
208	Functional Hazard Analysis	ENG	S	G	G	GC	GC
209	SoS Hazard Analysis	ENG	N/A	G	G	GC	GC
210	Environmental Hazard Analysis	ENG	S	G	G	G	GC
301	Safety Assessment Report	ENG	S	G	G	G	S
302	Hazard Management Assessment Report	ENG	S	G	G	G	S
303	Test and Evaluation Participation	ENG	G	G	G	G	S
304	Review of Engineering Change Proposals, Change Notices, Deficiency Reports, Mishaps, and Requests for Deviation/Waiver	ENG	N/A	S	G	G	G
401	Safety Verification	ENG	N/A	S	G	G	S
402	Explosives Hazard Classification Data	ENG	N/A	S	G	G	GC
403	Explosive Ordnance Disposal Data	ENG	N/A	S	G	G	S

Task Type
ENG – Engineering
MGT – Management

Program Phase
MSA – Materiel Solution Analysis
TD – Technology Development
EMD – Engineering and Manufacturing Development
P&D – Production and Deployment
O&S – Operations and Support

Applicability Codes
G – Generally Applicable
S – Selectively Applicable
GC – Generally Applicable to Design Change
N/A – Not Applicable

Summary

- The system safety process is:
  - Planned
  - Proactive
  - Formal
  - Forensic in nature
  - Closed-loop
  - Iterative
  - Validated (proven)
  - System oriented (integration of all subsystems)
  - Inclusive (HW, SW, Human, Env, Procedures)

Standards

DoD — MIL-STD-882E
System Safety, Department Of Defense Standard Practice, 2012 (Original 1969).

Commercial — ANSI/GEIA-STD-0010-2009
Standard Best Practices for System Safety Program Development and Execution, 2009.

Chapter 4 – System Safety Process

System Safety and Reliability Analysis
Course Notes

Chapter 5
System Safety Topics

Clifton A. Ericson II
Design Safety Solutions LLC
cliftonericson@verizon.net
540-786-3777

© C. A. Ericson II 2014

Tangible Benefits of System Safety

- Preventing loss of life and/or serious injury to personnel
- Preventing system loss
- Preventing harm to the environment and the surrounding community
- Saving dollars that might be lost from mishaps
- Maintaining a viable military force (by not losing systems, equipment and people to mishaps)
- Avoiding law suits and negative public opinion resulting from mishaps
- Meeting societal expectations for safety and risk

The results of system safety are often not visible because the SSP has been successful in preventing mishaps, and prevented mishaps are not a quantifiable metric. This phenomenon tends to under-rate and under-value the benefits of the SSP.

Chapter 5 – System Safety Topics

Cost of Safety

- Investment Costs
    - System Safety Program (SSP) costs
    - Software Safety Program costs
    - Special safety testing costs
    - Safety certification costs
    - Safety training costs
    - Special COTS safety program costs
    - Design change costs for safety enhancement
    - Independent safety assessment costs
    - Safety review boards and audit costs

- Mishap Loss Costs (Penalties)
    - Loss of life costs
    - Accident investigation costs
    - Legal costs
    - Loss of equipment costs
    - Equipment replacement costs
    - Equipment repair costs
    - Environmental repair costs
    - Loss of reputation costs
    - Loss of contracts costs
    - Political costs

> Safety is an investment

> If you think safety is expensive, try an accident!

Optimum Safety Cost

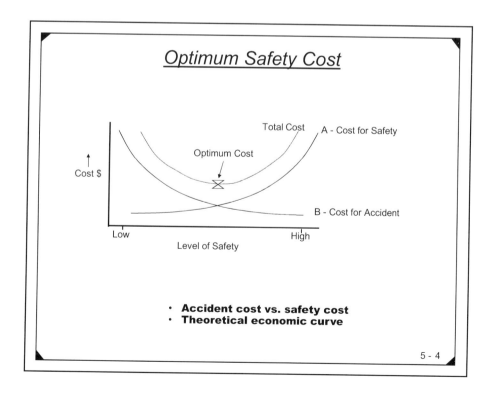

- **Accident cost vs. safety cost**
- **Theoretical economic curve**

Chapter 5 – System Safety Topics

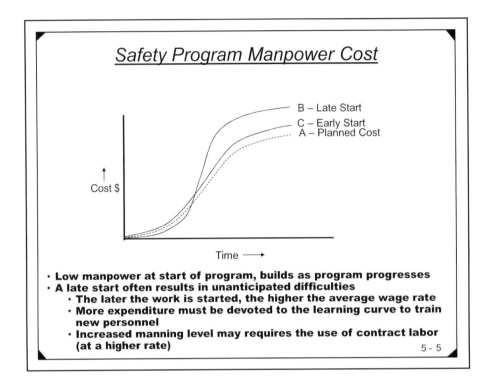

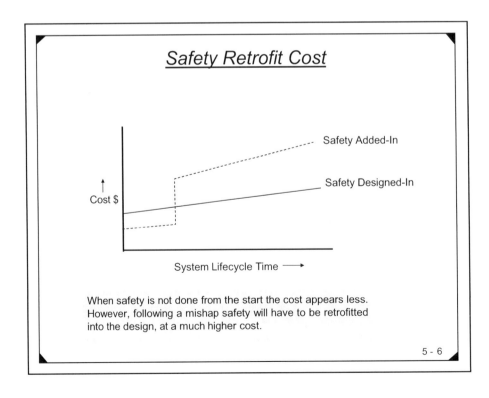

Chapter 5 – System Safety Topics

Cost of Safety

- The overall cost for a SSP should be commensurate with the mishap risk involved
- A high mishap risk program requires more safety effort
- Each system is unique in risk and safety cost

The cost of safety is an investment

5 - 7

Proactive Safety Costs Less

- Reduce Program Costs
    - Lower potential for major redesign costs
    - Reduces technical manual development
    - Reduces system training
- Reduce Schedule Risk
    - Lower potential for major redesign schedule delay
        - Design, manufacture, testing
- Reduce Operational Costs
    - Eliminate mishaps and mishap costs
    - More operational equipment available when needed

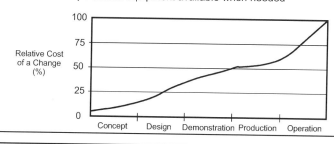

5 - 8

Chapter 5 – System Safety Topics

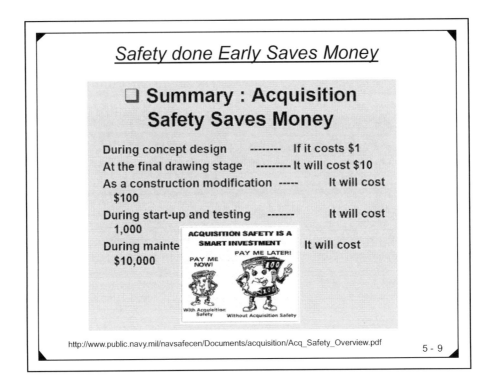

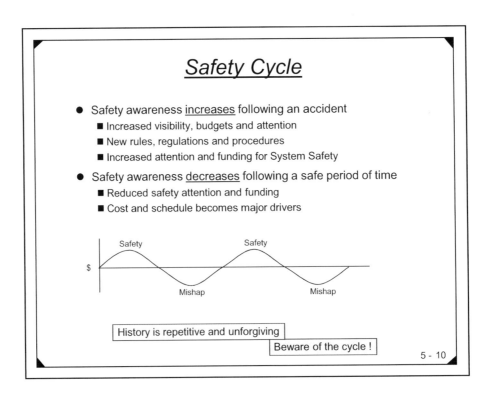

Chapter 5 – System Safety Topics

System Safety Program (SSP)

- A SSP is the combined set of <u>people</u> and tasks that <u>implement and execute</u> the system safety process on a development project or program.

- The SSP consists of an <u>organization</u> that performs the necessary system safety tasks and activities to realize the system safety objectives and obtain the necessary evidence of safety achievement.

In general, the overarching objectives of the SSP are to:

- Manage and execute the system safety process
- Perform the core system safety elements
- Ensure that system design meets applicable safety requirements
- Ensure that all system hazards are identified and controlled
- Develop a system presenting minimal mishap risk
- Protect the system and its users, the public and the environment from mishaps
- To intentionally design-in safety into the overall system design

SSP/SSPP Authority

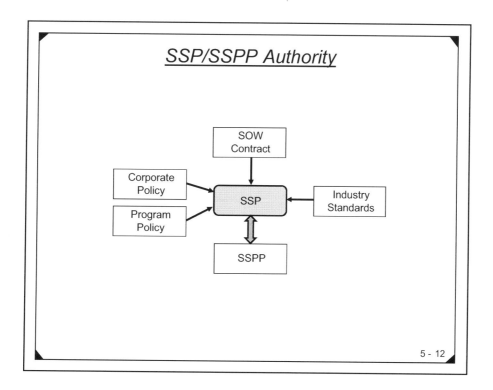

Chapter 5 – System Safety Topics

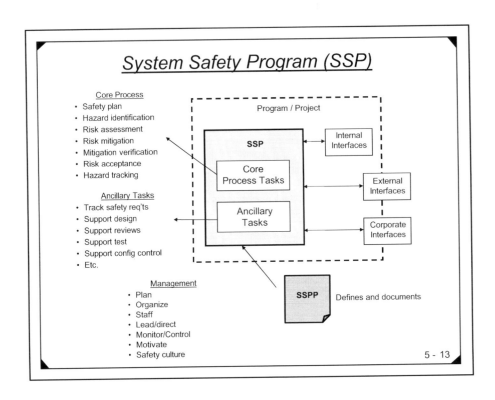

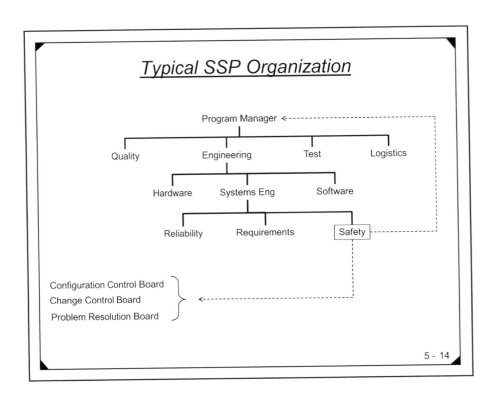

Chapter 5 – System Safety Topics

System Safety Program Plan (SSPP)

- The SSPP orchestrates the entire SSP
  - It defines the system safety program (SSP)
  - It outlines tasks, schedules, methods, risk criteria, etc.
  - It defines risk criteria and risk approval process
  - It provides mgt and customer understanding and buy-in
  - It is a "living" document (It can be updated and changed during the life of the program)

SSPP Contents

- A brief system overview description, including the identification of subsystems, functions and modes of operation and maintenance
- The SSP policies (Goals, Guidelines, Precepts, Definitions, etc.)
- Defines the system safety organization and lines of communication
- Identify the Government PFS, who serves as the PM's system safety agent
- Establish the SSWG Charter and set forth a plan of action for the SSWG
- Define the approach and methodologies for:
  - Hazard identification
  - Risk management (assessment, mitigation, verification and acceptance)
  - Hazard tracking
  - Software safety
  - COTS/NDI safety
  - HSI safety
  - Critical Safety Item (CSI) identification

Chapter 5 – System Safety Topics

SSPP (continued)

- Define the system safety HRI matrix for the program
- Define the software safety criticality matrix and the level of rigor tasks
- Define the program's formal hazard closure and residual risk acceptance process
- Define the Safety Assessment Report (SAR) format and schedule
- Define the interface support program (internal and external to the program)
- Define Technical Specialty programs (Laser, EMI, Ordnance, Electrical, etc.)
- Define the Safety Audit Process
- Define the Critical Safety Item (CSI) Identification program
- Define the Failure and Mishap Analysis And Reporting Process
- Define the program safety products and their schedule for completion
- Define the overall safety tasks, schedules and milestones
- Define safety responsibilities and authority
- Define the required resources and personnel qualifications
- Define the safety analysis methodologies and formats to be utilized (e.g., PHA, FTA)

SSPP Document Outline

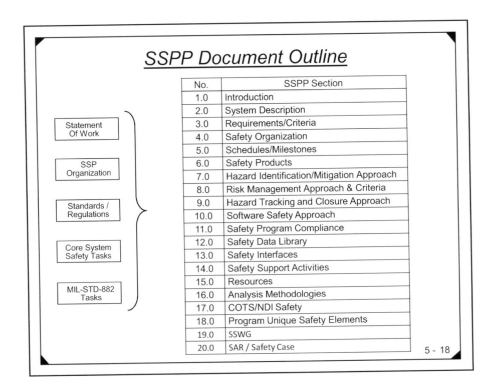

No.	SSPP Section
1.0	Introduction
2.0	System Description
3.0	Requirements/Criteria
4.0	Safety Organization
5.0	Schedules/Milestones
6.0	Safety Products
7.0	Hazard Identification/Mitigation Approach
8.0	Risk Management Approach & Criteria
9.0	Hazard Tracking and Closure Approach
10.0	Software Safety Approach
11.0	Safety Program Compliance
12.0	Safety Data Library
13.0	Safety Interfaces
14.0	Safety Support Activities
15.0	Resources
16.0	Analysis Methodologies
17.0	COTS/NDI Safety
18.0	Program Unique Safety Elements
19.0	SSWG
20.0	SAR / Safety Case

Chapter 5 – System Safety Topics

System Safety Working Group (SSWG)

- The System Safety Working Group (SSWG) is an integrated team of safety and non-safety personnel whose continuous activities monitor and guide the SSP throughout the program lifecycle.
- The SSWG is responsible for ensuring that safety efforts are closely coordinated and safety tasks are completed on an effective and timely basis.
- The SSWG is a Government conducted safety activity. The PFS serves as the Chairman of the SSWG.
- SSWG membership is comprised of experienced Government and Contractor personnel involved in the development, installation, and life-cycle support of the system.
- SSWG meetings will be held on a regular basis (not less than once a year) to assess system safety issues.
- The SSWG reviews and reports on safety program status, resolves safety issues, and forwards recommendations to the PM. At a minimum, all Open and Monitor hazards will be reviewed at each SSWG. Newly proposed hazards will also be reviewed at SSWG meetings. Subject matter experts (SMEs) will brief the SSWG on the technical aspects of safety issues.
- The PFS is responsible for establishing the SSWG, the SSWG charter and conducting regularly scheduled SSWG meetings. The PFS is also responsible for ensuring that SSWG meeting minutes are documented and all SSWG action items are resolved on a timely basis.

SSWG Functions

- Review of proposed hazards for credibility and induction into the HTS
- Review and concurrence on the HRI of identified hazards
- Review and concurrence on the Open, Monitor or Close status of hazards
- Improve communication between all program safety participants
- Provide for the exchange of safety information
- Assemble pertinent safety data to aid management decisions
- Coordinate safety issues between interfacing contractors
- Discuss the resolution of identified hazards
- Collect and review data as necessary, such as accident/incident reports
- Ensure safety action items and issues are resolved and closed

Chapter 5 – System Safety Topics

SSWG Charter Outline

- Purpose
- Membership composition
- Chairperson
- Meeting rules
- Scope of SSWG
- Tasks, roles and responsibilities
- Meeting minutes (responsibility, maintaining a minutes file)
- Action Item list (responsibility, tracking, closure)
- Documenting and distributing meeting agendas and minutes
- Meeting and location schedule
- The process for reviewing, tracking and closing hazards
- New business (new safety issues and/or concerns)

The SSWG charter, membership and schedule should be documented in the SSPP

Government PFS Organization

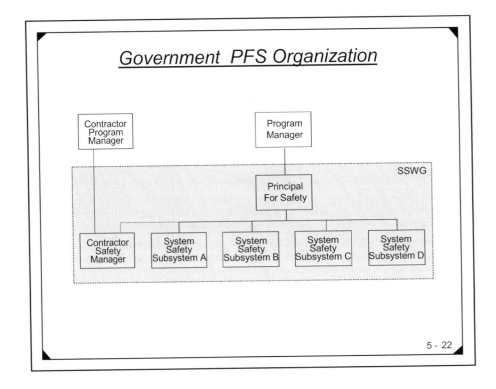

Chapter 5 – System Safety Topics

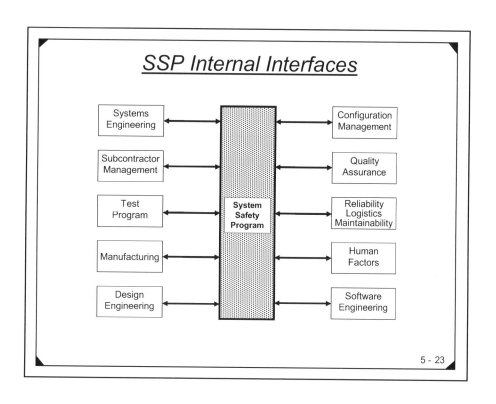

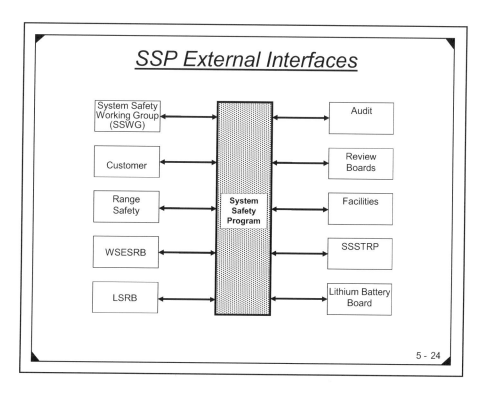

Chapter 5 – System Safety Topics

Safety Audit

- To conduct an independent examination of the project
  - Ensure the SSP is being performed and managed effectively
  - Ensure the SSP is meeting contractual requirements
  - Ensure that relevant safety aspects have not been omitted or overlooked
  - Ensure thoroughness and completeness in the SSP

- To disclose strengths, weaknesses and possible problems in the SSP
  - Provide management assurance
  - Allow for SSP adjustments before its essentially too late

- Customers
  - The company
  - The program manager
  - The SSP manager
  - The customer

Audit Factors

- The main audit types include:
  - Process performance
    - Ensure contractual and standard best practices are begin followed
    - e.g., hazards are properly closed
  - Physical
    - To check physical product parameters against requirements
    - e.g., sharp edges
  - Process improvement
    - To find areas in process improvement
    - e.g., manufacturing

- Auditor requirements:
  - Auditors must be independent from program
  - Auditors have the requisite skills, knowledge and experience
  - Auditors must be able to apply diplomacy

Chapter 5 – System Safety Topics

General Audit Model

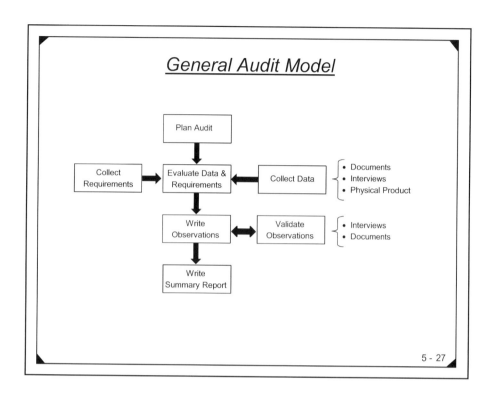

Audit Review Areas

- System Safety Program (SSP) organization
- Program and safety milestones
- System Safety Program Plan (SSPP)
- Safety requirements
- Hazard analyses
- Risk criteria and risk assessment
- Hazard mitigation (risk resolution)
- Hazard tracking, acceptance and closure
- Safety verification and validation
- Change proposal process
- Software safety
- Safety staffing / funding
- Management and program safety culture

Considerations:
- Process
- Requirements
- Evidence
- Quality
- Completeness
- Correctness

Chapter 5 – System Safety Topics

Pricing a SSP

- A basic element of system safety is pricing the SSP program
  - When building a system/product for a customer, the SSP cost must be included in the total cost proposal and contract
  - When building a system/product for general sales, the SSP cost must be estimated and factored into the overall cost

- Pricing the SSP is just one critical aspect of contracting

SSP Pricing/Contracting

- Contracting is the function or process that actually consummates the system acquisition process
- It provides the legal framework between the purchasing agency (i.e., the Developer) and the producing agency, (i.e., the Contractor)
- The contracting process also formulates all of the system requirements, including system safety, into the contract
- Only the items specified in the contract will be performed; therefore it is essential that contracting be done correctly and completely from the start
- After the contract is agreed upon, any changes the Developer wants to make must be done in a separate negotiation and contract modification, with additional funding provided

- The SSPP is key to pricing a SSP
  - All of the tasks, schedules and resources requirements are listed in the SSPP
  - These items can be individually priced
  - Thus, it is imperative that a complete SSPP be established
  - The SSPP is part of the final contract

Chapter 5 – System Safety Topics

Developer's Responsibility

- Contractors will provide a much better response when they know exactly what the safety requirements are without having to second-guess what the Developer intended.
- For this reason the Developer's safety representative should have expertise in system safety to ensure the correct safety requirements are included.
- The RFP, prepared by the Developer, must address the following system safety aspects as a minimum:
  - Definitive system safety objectives and goals
  - Definitive system safety program requirements
  - Required safety data (CDRL and non-deliverable Contractor generated data)
  - System Safety Program Plan (SSPP) requirements
  - Mishap risk categorization criteria for hazards and hazard acceptance criteria
  - Historical safety data as available
  - Technical data on GFE/GFP

Generic Contracting Process

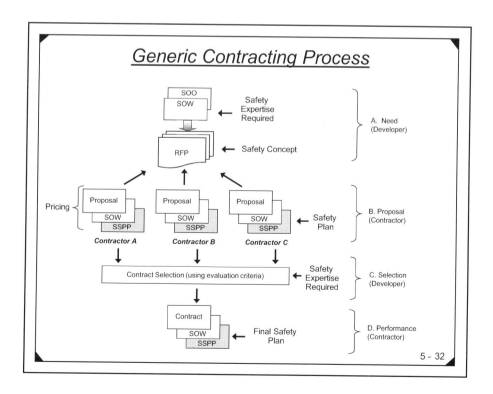

Chapter 5 – System Safety Topics

COTS?

- Products that encompass a wide variety of general-purpose off-the-shelf (OTS) products
- A previously developed item (hardware or software)
- Purchased OTS or is reused from another application or program

- Commercial Off-The-Shelf (COTS)
- Non Developmental Items (NDI)
- Previously Developed Software (PDS)
- Government Furnished Equipment (GFE)

COTS Advantages/Disadvantages

Advantages
- Cost savings (no development costs)
- Rapid insertion of new technology
- Proven product/process
- Possible broad user base
- Technical support
- Logistics support

Disadvantages
- Unknown development history (standards, quality assurance, test, analysis, failure history)
- Design and test data unavailable (drawings, test results, etc.)
- Proprietary design (prohibits information)
- Unable to modify
- Unknown limitations (operational, environmental, stress, etc.)
- May not be supported (configuration control, tech support, updates, etc.)
- Many include extra unnecessary capabilities
- Unknown part obsolescence factor
- Safety analyses unavailable or not applicable
- May require increased test and analysis for safety verification

Chapter 5 – System Safety Topics

Concerns with COTS

- Outside of any control boundary
- Minimal or no access to the source code or schematics
- No guarantee of component evolution
- Maintainability issues
- Update issues
- Ownership of data issues
- Functional limitations
- Unknown functions

COTS Safety Process

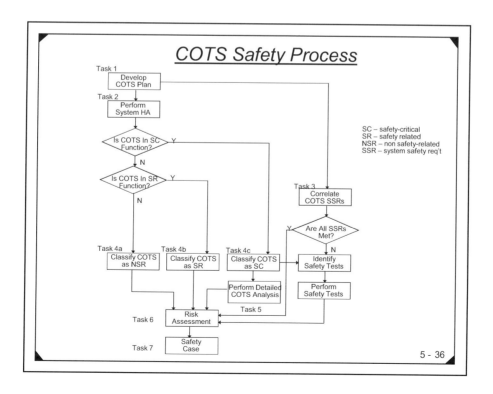

Chapter 5 – System Safety Topics

COTS

- COTS can present many problems for safety, particularly when used in safety-critical functions or applications
- It is recommended that COTS be evaluated for safety impact via hazard analysis and system testing
- The use of COTS is not an excuse to avoid performing safety and hazard analysis on the COTS items in the new system design and environment
- COTS use *DOES NOT* relinquish safety responsibility, thus unexpected additional costs may be incurred in order to provide safety evidence for safe use of COTS
- Make sure COTS is the right solution

Safety Assessment Report (SAR)

- The SAR provides an assessment of the identified hazards and the residual mishap risk.
- The SAR is a snapshot of program safety coverage and risk at a point in time. A SAR is generated several times during the life of a program, to assist in evaluating each program development milestone.
- The SAR is a decision tool for organizations inside and outside the program and is used to determine if program mishap risk is acceptable, and if the SSP is on track.
- The SAR is intended to provide an argument and evidence that the system is safe for use
- The SAR is a comprehensive evaluation of the safety risks being assumed prior to test or operation of the system or at contract completion.
- A SAR is not a hazard analysis, but an assessment of the SSP system program mishap risk derived from the safety and hazard analyses.

Chapter 5 – System Safety Topics

Typical SAR Content

- Purpose of the SAR
- System description (including hardware, software and COTS)
- SSP description
- Hazard analysis tasks performed
- Hazard assessment
- Risk assessment
- Top Level Mishap (TLM) summary
- List of design safety features
- Hazardous materials assessment
- Conclusions
- Recommendations
- Operational Limitations
- Safety Statement
- Appendix (detailed hazard analyses and source documents)

Safety Case

As Safety Case is:

- A safety case is a formal documented body of evidence that provides a convincing and valid argument that a system is adequately safe for a given application in a given environment
- Elements of the safety case include safety claims, evidence, arguments, and inferences

A Safety Case includes:

- Claims – statements or requirements about a safety property of the system, subsystem or component.
- Evidence – Information that demonstrates the validity of the claim. This can include facts (e.g., based on observations or established scientific principles), analysis, research conclusions, test data, or expert opinions.
- Arguments – the logical and rational link between safety claims and the evidence supporting the claims; logical information linking the evidence to the claim, where inference is typically the mechanism used.

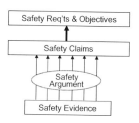

SAR alternative.

Chapter 5 – System Safety Topics

Safety Culture

- The safety culture of an organization is the resulting product of the individual and group values, attitudes, competencies and patterns of behavior that establish the organization's commitment to safety.
- Organizations with a positive safety culture are characterized by communications founded on trust, by shared perceptions of the importance of safety and by confidence in the effectiveness of preventive safety measures.
- A positive safety culture establishes an atmosphere where the employees are fully dedicated to safety and system safety and they receive management support in this endeavor.
- In a negative safety culture the opposite is true; safety commitment is strangled by the cynicism of some individuals, there is a fear of expending too much effort on safety, and there is a visible lack of management support.
- Safety is most effectively achieved when management makes safety a core value of the company. Leadership and commitment from the chief executive are required. Safety attitude and commitment flows from the top down, not the bottom up.
- It has been demonstrated that once safety becomes a corporate "habit" other corporate tasks are performed more effectively and profits increase.

Summary

- System safety is predicated on the concept that the most effective means to avoid mishap during system operation is by eliminating or reducing hazard risk during design and development
- Cost accounting does not show costs saved through accident prevention, therefore safety is often difficult to visualize and justify
- Correctly establishing and pricing the SSP is important
    - The Contractor can do no more than priced and contracted
    - Contract changes require additional funding
    - Past experience cost data can aid in estimating new SSPs
- Human error mishaps are typically really design errors
- Safety is no accident, it must be created (via system design)

Chapter 5 – System Safety Topics

System Safety and Reliability Analysis
Course Notes

Chapter 6
Reliability

Clifton A. Ericson II
Design Safety Solutions LLC
cliftonericson@verizon.net
540-786-3777

© C. A. Ericson II 2014

Reliability

- Reliability deals with the task of ensuring that a system performs a required task or mission for a specific period of time.

- System safety deals with the task of ensuring that a mishap does not occur within a system during system operation or standby.

- Availability deals with the task of ensuring that a system is ready for use when needed (assumes repair is possible).
  - Downtime results from failures and mishaps (R & S)

Chapter 6 – Reliability

Reliability Definition

- **Reliability** is the probability that a system will not fail for a given period of time under specified operating conditions.
    - Reliability is an inherent system design characteristic
    - Reliability plays a key role in determining the system's cost-effectiveness and availability for use

- **Reliability** engineering is a specialty discipline within the systems engineering process; key activities include:
    - *Design* - including design features that ensure the system can perform in the predicted physical environment throughout the mission
        - Fix problems in design rather than in the field
    - *Trade studies* - reliability as a figure of merit; often traded with cost
    - *Modeling* - reliability models predict system reliability
    - *Test* – provide independent predictions of system reliability

- Reliability is an emergent property (as is safety)
- Reliability applies to: system and system components, such as electrical, mechanical, software, human

6 - 3

Reliability Basics

- Reliability is a time dependent characteristic
    - It can only be determined after an elapsed time but can be predicted at any time

- Reliability cannot be tested into a product
    - It must be designed and manufactured into it
    - Testing only indicates how much reliability is in the product

6 - 4

Chapter 6 – Reliability

Reliability Tasks

- Reliability program plan
- Establish reliability requirements for system
- Designing for reliability (redundancy, high rel parts, etc.)
- Reliability modeling
  - Allocations
  - Prediction (RBDs, FTA, FMEA, test)
  - Actuals
- Supporting program design reviews
- Monitor and control of subcontractors & suppliers
- Failure reporting, analysis, corrective action system (FRACAS)
- Reliability testing
- Reliability monitoring (development, field)
- Data collection

Reliability Lifecycle Activities

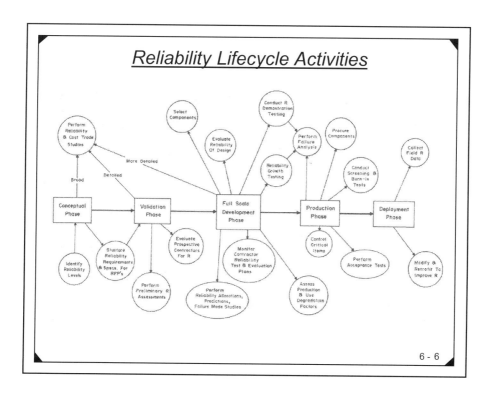

Chapter 6 – Reliability

Reliability Tasks

Reliability Tasks	Life Cycle Phase				
	Conceptual	Validation	Full Scale Development	Production	Deployment
Requirements Definition	xxxxxxxxx	xxxxxxxAAAAAA	....		
Reliability Model		xxxxxxxxxxxxxxxxxxxxx	......		
Reliability Prediction		xxxxxxxxxxxxxxxxxxxxx	......		
Reliability Apportionment		ooooopoooooooooooooooo	......		
Failure Modes Analysis		ooooopoooooooooooxxxxx	......		
Design for Reliability		ooooopxxxxxxxxxxxxxxxxxxxxx	....		
Parts Selection		ooooopxxxxxxxxxAAAAAA	......		
Design Review		ooooopxxxxxxxxxxxxxxxxx	......		
Design Specifications	xxxxxxx	xxxxxxxxxxxxxx	....		
Acceptance Specifications		xxxxxxxxxxAAAAA	....		
Reliability Evaluation Tests		---- xxxxxxxxxxxxxxxxx			
Failure Analysis		---- xxxxxxxxxxxxxxxxxxxxxxxxx	ooooooooooooo	oooooooooo	
Data System		---- xxxxxxxxxxxxxxxxxxxxxxxxx	ooooooooooooo	oooooooooo	
Quality Control		ooooooooooxxxxxxxxxxxxxx	xxxxxxxxxxx	ooooooooooo	
Environmental Tests		xxxxx	......AAAAAAA	....	
Reliability Acceptance Tests		xx	.......AAAAAoooooooooo		

Legend:
- 00 Necessary
- xx Very Important
- AA Critical
- .... Low key
- ---- Desirable

Reliability Methods

- Qualitative Reliability
  - Any failure mode that is eliminated improves reliability
  - Amount of improvement is not known
  - Identify failure modes and eliminate/reduce
  - Provide corrective action for actual failures

- Quantitative Reliability
  - Numbers game
  - Predict and measure reliability
    - Original design
    - Design changes

Chapter 6 – Reliability

Basic Reliability Equations

- $R = e^{-\lambda T}$
- $R + Q = 1$
- $Q = 1 - R = 1 - e^{-\lambda T}$
  - $Q \approx \lambda T$ when $\lambda T < 0.001$ (approximation)
- Where:
  - R = Reliability or <u>Probability of Success</u>
  - Q = Unreliability or <u>Probability of Failure</u>
  - λ = component failure rate = 1 / MTBF
  - MTBF = mean time between failure
  - T = time interval (mission time or exposure time)

Measures of Reliability

Mean Time To Failure (MTTF)
- The average time that elapses until a failure occurs
- For non-repairable systems
- MTBF = Number of operational hours / Number of failures
- $\lambda = 1.0 / MTTF$

Mean Time Between Failure (MTBF)
- The average time between successive failures
- For repairable systems
- MTBF = Number of operational hours / Number of failures
- $\lambda = 1.0 / MTBF$

TTF – Time to failure
TBF = Time between failure
TTR = Time to repair

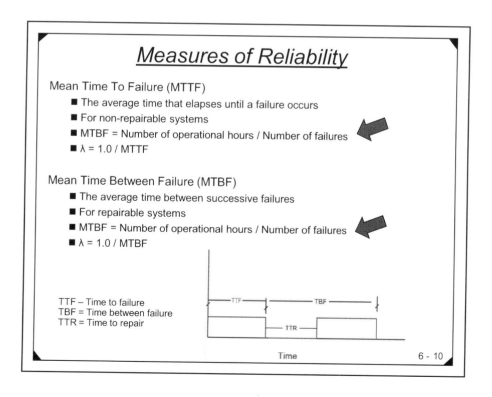

Chapter 6 – Reliability

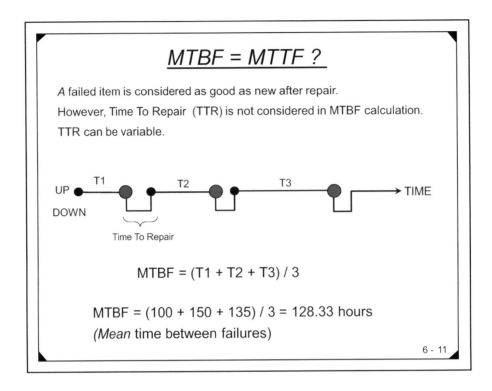

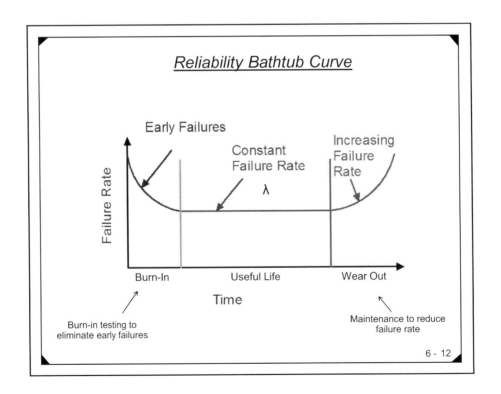

Chapter 6 – Reliability

The Bathtub Curve

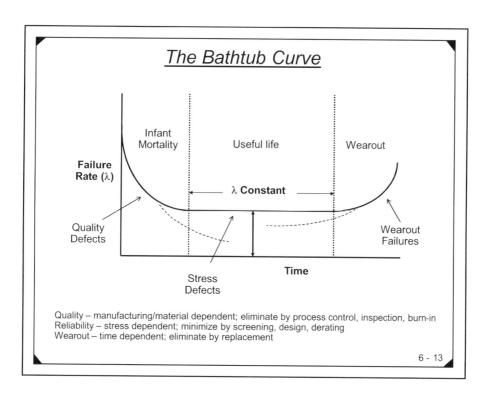

Quality – manufacturing/material dependent; eliminate by process control, inspection, burn-in
Reliability – stress dependent; minimize by screening, design, derating
Wearout – time dependent; eliminate by replacement

Typical Failure Causes

Infant Mortality Failure Causes	Useful Life Failure Causes	Wearout Failure Causes
• Workmanship • Quality control • Substandard materials • Contamination • Human error • Assembly problems • Poor processes • Poor design	• Stress • Low safety factors • Defects • Abuse • Misuse • Human error • Random failures • Design exceeded	• Wear • Ageing • Fatigue • Creep • Short design life • Poor maintenance • Degradation

Chapter 6 – Reliability

Value of Reliability Models

- Allocations
  - Allocate R goals to subsystems and components
  - During early development
- Prediction (RBDs, FTA, FMEA, test)
  - During mid development
  - Predict expected R values
- Actuals
  - During mid-final development
  - Determine actual R values being achieved

- The earlier it is determined that allocated goals are not being achieved the earlier corrective changes can be made

Reliability control is a powerful reliability design method

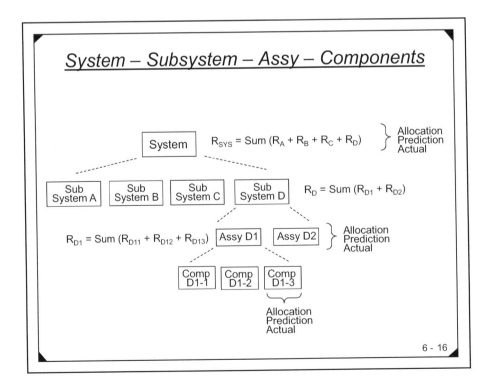

Chapter 6 – Reliability

Allocations & Predictions

System Goal: Computer R=0.95

Allocations (System Total = 0.95)

| Motherboard R=0.99 | HDD R=0.99 | Pwr Supply R=0.99 | Monitor R=0.99 | Keyboard R=0.99 |

Predictions (System Total = 0.93) << Not going to meet goal or requirement

| Motherboard R=0.96 | HDD R=0.98 | Pwr Supply R=0.999 | Monitor R=0.99 | Keyboard R=0.999 |

Prediction estimates are from data, testing

Reliability control is a powerful reliability design method

Reliability Prediction

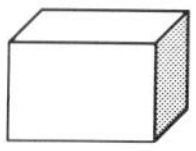

Radio Transmitter

Component	Qty	FR Each	FR Sum
Transistors	20	0.012	0.240
Resistors	121	0.001	0.121
Diodes	35	0.002	0.070
Capacitors	15	0.002	0.030
Switches	5	0.008	0.040

Total FR = 0.501

FR = λ = failures per million hours (fpmh)

Chapter 6 – Reliability

Reliability Block Diagram (RBD)

- A RBD is a graphical representation of the components of the system and how they are reliability-wise related (connected)

- Prediction model

Series Reliability

In a series configuration, a failure of any component results in failure for the entire system; all of the units in a series system must succeed for the system to succeed.

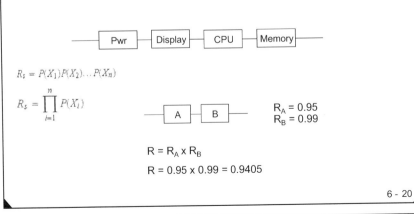

$$R_s = P(X_1)P(X_2)\ldots P(X_n)$$

$$R_s = \prod_{i=1}^{n} P(X_i)$$

$R_A = 0.95$
$R_B = 0.99$

$R = R_A \times R_B$
$R = 0.95 \times 0.99 = 0.9405$

Chapter 6 – Reliability

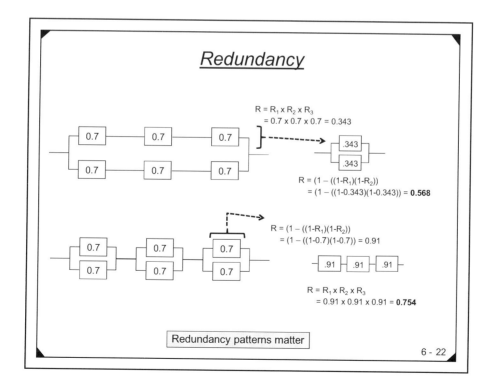

Chapter 6 – Reliability

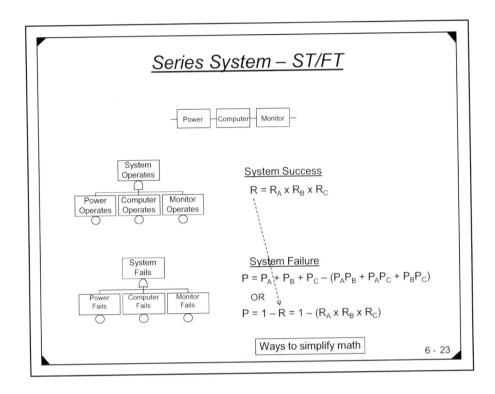

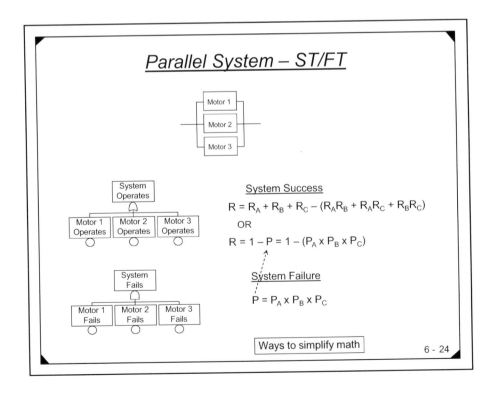

Chapter 6 – Reliability

Effects of Failure Rate & Time

- The longer the mission (or exposure time) the higher the probability of failure

- The smaller the failure rate the lower the probability of failure

- It's important to control either the operational time or the failure rate

> The Effect of Exposure Time on Probability is Significant

> Time and stress are system enemies

The Effect of Time on Failure Probability

Time is your enemy!

$\lambda = 1.0 \times 10^{-6}$

λ_A (Hrs)	T (Hrs)	P_A
1.0 x 10-6	1	9.99E-7
1.0 x 10-6	10	9.99E-6
1.0 x 10-6	100	9.99E-5
1.0 x 10-6	1,000	9.99E-4
1.0 x 10-6	10,000	9.99E-3
1.0 x 10-6	100,000	0.095
1.0 x 10-6	1,000,000	0.6321
1.0 x 10-6	10,000,000	0.99995

$$P_A = 1 - e^{-\lambda T}$$

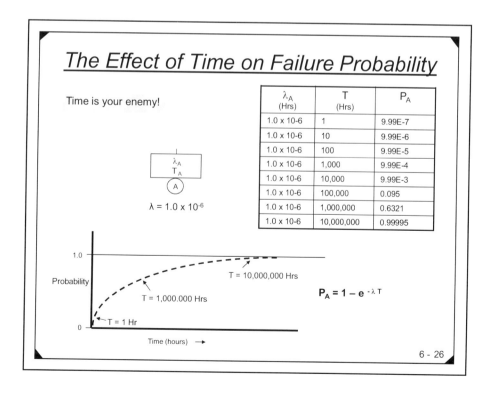

Chapter 6 – Reliability

The Effect of Failure Rate On Probability

This is why small λ is important

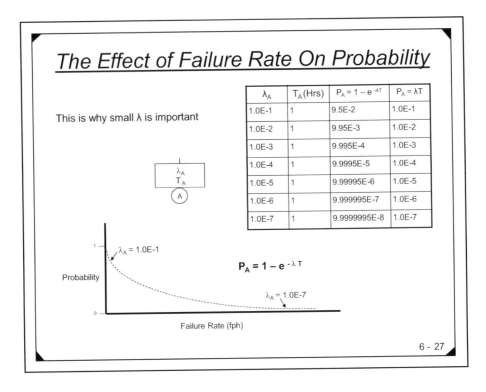

λ_A	T_A(Hrs)	$P_A = 1 - e^{-\lambda T}$	$P_A = \lambda T$
1.0E-1	1	9.5E-2	1.0E-1
1.0E-2	1	9.95E-3	1.0E-2
1.0E-3	1	9.995E-4	1.0E-3
1.0E-4	1	9.9995E-5	1.0E-4
1.0E-5	1	9.99995E-6	1.0E-5
1.0E-6	1	9.999995E-7	1.0E-6
1.0E-7	1	9.9999995E-8	1.0E-7

$$P_A = 1 - e^{-\lambda T}$$

6 - 27

Testing

- RDT – Reliability Demonstration Testing
    - Demonstrate the steady state reliability of a product (system level)
    - Demonstrate MTBF

- ALT – Accelerated Life Testing
    - Discover the useful life of a product in a short time
    - Identify expected failure modes
    - Subject product to continuous operation (24/7)
    - Demonstrate wearout

- HALT – Highly Accelerated Life Testing
    - Discover product weakness and design margins
    - Subject product to stimuli well beyond expected environments
    - Stresses – hot, cold, vibration, cycles, combined

6 - 28

Chapter 6 – Reliability

Testing

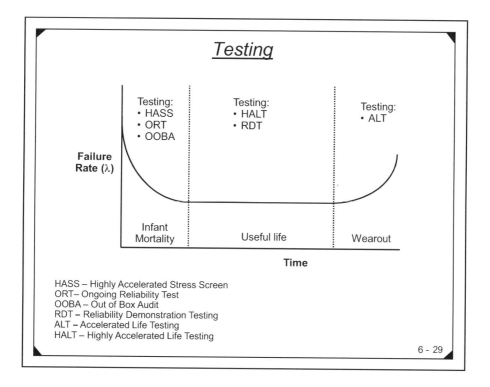

HASS – Highly Accelerated Stress Screen
ORT – Ongoing Reliability Test
OOBA – Out of Box Audit
RDT – Reliability Demonstration Testing
ALT – Accelerated Life Testing
HALT – Highly Accelerated Life Testing

Reliability Design Features

- Redundancy
- Derating (i.e., safety margin)
- Use of high reliability components
- Failure rate control (of parts, components, black boxes, etc.)
- Reliability control of black boxes
- Exposure time control
- Fault tolerance
- Fault detection and correction
- High quality manufacturing
- Increase mean strength (size, weight, material strength, etc.)
- Decrease stress
- Use high safety margin (strength vs. Stress)

Chapter 6 – Reliability

Maintainability (M)

- The probability of performing a successful repair action within a given time
- If a component has a 90% maintainability in one hour, then there is a 90% probability that the component will be repaired within an hour
- M affects reliability, availability, downtime, cost of operation
- Maintenance actions can be divided into three types: corrective maintenance, preventive maintenance and inspections
- Involves designing specifically for maintainability

$$M(t) = 1 - e^{-\mu t}$$

μ = repair rate

$$\frac{1}{\mu} = MTTR \text{ (mean time to repair)}$$

Availability (A)

- The probability that the system is operational when it is requested for use.
- The probability that a system is not failed or undergoing a repair action when it needs to be used.
- Availability is a performance criterion for repairable systems that accounts for both the reliability and maintainability properties of a component or system.
- For example, if a lamp has a 99.9% availability, there will be one time out of a thousand that someone needs to use the lamp and finds out that the lamp is not operational either because the lamp is burned out or the lamp is in the process of being replaced.

$$A_I = \frac{MTBF}{MTBF + MTTR}$$

Chapter 6 – Reliability

Reliability, Availability and Maintainability

- Reliability = MTBF/(1+MTBF)

- Availability = MTBF/(MTBF + MTTR)

- Maintainability = 1/(1 + MTTR)

6 - 33

Quantitative Failure Data

- FTA and HA Requires failure data
- Finding qualified data can be difficult
- Higher confidence data provides higher confidence results

- However, low confidence data can still provide useful results
    - If top probability is acceptable when using worst case estimates, then more refined data may not be necessary

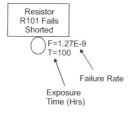

6 - 34

Chapter 6 – Reliability

Finding Failure Data

- High confidence failure data provides high confidence results
  - Engineering judgment can be useful however
- Collect and file failure data whenever you encounter it
- Some commercial data sources are available
- Allow research time for finding adequate data
- Testing provides data
  - ALT – accelerated life testing
  - HALT – highly accelerated life testing

Component Failure Rate Sources

- Electrical
  - MIL-HDBK-217
- Mechanical
  - NSWCSTD_98-LE1 (Handbook of Reliability Prediction Procedures for Mechanical Equipment)
  - WASH-1400 Reactor Safety Study, 1975
- Human
  - Gertman, David I. & Blackman, Harold S., *Human Reliability & Safety Analysis Data Handbook*. John Wiley & Sons Inc., New York, NY, 1984
  - WASH-1400 Reactor Safety Study, 1975
- Other
  - Telcordia (Bellcore)
  - PRISM
  - 217Plus
  - Chinese GJB/z 299B/C
  - IEC TR 62380
  - Siemens SN29500
  - HRD5

Chapter 6 – Reliability

SRC Data

- Alion System Reliability Center
  - http://src.alionscience.com

- ERRD-CD – Electronic Parts Reliability Data, $395, 1997
  - SRC-STD-6100: contains reliability data on both commercial and military electronic components for use in reliability analyses. It contains failure rate data on integrated circuits, discrete semiconductors (diodes, transistors, optoelectronic devices), resistors, capacitors, etc.

- FMD-97 - Failure Mode/Mechanism Distributions, $120, 1998
  - SRC-STD-6300: a cumulative compendium of failure mode and mechanism data. Quantifies modes and mechanisms by providing their relative probabilities of occurrence.

RIAC Data

- Reliability Information Analysis Center
  - http://www.theriac.org
  - Sponsored by DoD

- Electronic Parts Reliability Data, $295

- Failure Mode/Mechanism Distributions, $75

- Software Reliability Sourcebook, $60

Chapter 6 – Reliability

Human Error Rates

Typical Human Operator Failure Rates...

ACTIVITY	ERROR RATE
* Error of omission / item embedded in procedure	3×10^{-3}
* Simple arithmetic error with self-checking	3×10^{-2}
* Inspector oversight of operator error	10^{-1}
* General rate / high stress / dangerous activity	$0.2 - 0.3$
** Checkoff provision improperly used	$0.1 - 0.9$ (0.5 avg.)
** Error of omission / 10-Item check list	$10^{-4} - 5 \times 10^{-3}$ (1×10^{-3} avg.)
** Carry out plant policy / no check on operator	$5 \times 10^{-3} - 5 \times 10^{-2}$ (5×10^{-2} avg.)
** Select wrong control / group of identical, labeled, controls	$10^{-3} - 10^{-2}$ (3×10^{-3} ave.)

Sources:
* WASH-1400 (NUREG-75/014); "Reactor Safety Study — An Assessment of Accident Risks in U.S. Commercial Nuclear Power Plants," 1975 (See Appendix I, attached.)
** NUREG/CR-1278; "Handbook of Human Reliability Analysis with Emphasis on Nuclear Power Plant Applications," 1980

Note: Similar data appear for similar activities in "Loss Prevention in the Process Industries"; 2nd Edition; F. P. Lees; Butterworths; 1996; ISBN 0 7506 1547 8.

Ref: Human Factors and Operator Errors, P.L. Clemens, 2002

Human Error Rates

Mismatches Abound...
between analytical predictions and experience:

- WASH-1400 — Human error probability for routine repetitive tasks:
  3×10^{-3} to 1×10^{-2} per individual operation.*

- Concert pianist, performing K.453, 1st movement:
  3996 individual, critical keystrokes.**

- Expected errors per performance:
  $3996 \times (3 \times 10^{-3}) \approx 12$ to $3996 \times (1 \times 10^{-2}) \approx$ **40 errors**
  A DOOMED REPUTATION!

SOURCES:
* WASH-1400 (NUREG-75/014); "Reactor Safety Study — An Assessment of Accident Risks in U.S. Commercial Nuclear Power Plants," 1975.
** Mozart, W. A.; 17th Piano Concerto (incl. Composer's own first cadenza).

Ref: Human Factors and Operator Errors, P.L. Clemens, 2002

Chapter 6 – Reliability

Military Data Centers

- Navy Data Center
- Air Force Data Center

- Track failures, mishaps and hours of operation for various systems

- Remember
  - λ = failure rate
  - λ = Number of failures / Total hours of operation
  - Example:
    - λ = 2 failures / 4,300,000 hours
    - = 0.000000465 failures / hour
    - = 4.65×10^{-7} failures / hour

Summary

- Reliability deals with the task of ensuring that a system performs a required task or mission for a specific time.
  - R = probability that system successfully operates for specified period of time in specified environment

- System safety deals with the task of ensuring that a mishap does not occur within the system while performing its task or mission.
  - S = acceptable hazard risk

- In general, a more reliable system is a safer system and a safer system is a more reliable system, however, there are exceptions.

Chapter 6 – Reliability

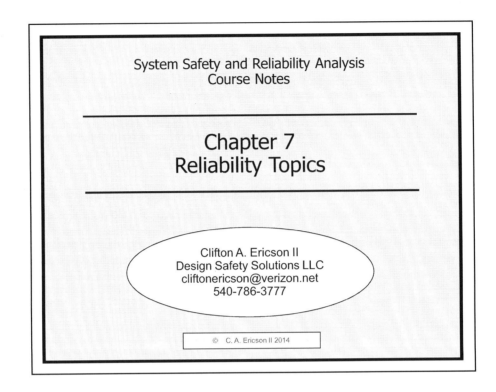

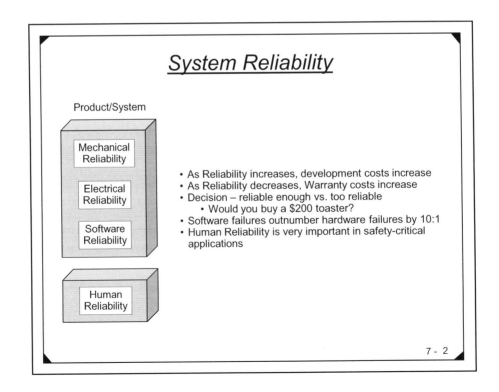

Chapter 7 – Reliability Topics

Reliability Balancing Act

- Three philosophies for a Reliability program
  - Build, test, fix – less time on analysis, more time on testing
  - Analytical – more time on analyses and less time on testing
  - Balanced – a balance between analysis and testing

- Two approaches for a Reliability program
  - Reliability measurement – uses R techniques to measure actuals vs. goals
    - R predictions
    - Accelerated Life Tests (ALT)
    - R Demonstration Tests (RDT)
    - Ongoing R Tests (ORT)
  - Reliability improvement – use R techniques to designed to improve product
    - Highly Accelerated Life Tests (HALT)
    - FMEA

Reliability Can Degrade Safety & Vice Versa

Hazard = Inadvertent launch when switch (SW) <u>fails closed</u>

Pwr → [SW1] → Missile

$R = 0.9$ (switch works correctly)   **R**
$P_H = 1 \times 10^{-6}$ (switch fails closed)   **S**

Pwr → [SW1 / SW2] → Missile (parallel)

$R = (1 - (1-R_1)(1-R_2)) = 0.99$ **R+**
$P_H = (1 \times 10^{-6}) + (1 \times 10^{-6}) = 2 \times 10^{-6}$ **S-**

Pwr → [SW1] → [SW2] → Missile (series)

$R = R_1 \times R_2 = 0.81$ **R-**
$P_H = (1 \times 10^{-6}) \bullet (1 \times 10^{-6}) = 2 \times 10^{-12}$ **S+**

| Safety / reliability tradeoff |

Chapter 7 – Reliability Topics

Reliability - Safety

- Safety is typically missing from the Reliability equation
- Reliability definition should be expanded
    - $R = \Sigma (R_{Failures})$ [old model]
    - $R = \Sigma (R_{Failures} + S_{Failures})$ [new model]
- Examples
    - Faults cause aircraft fire, which in turn causes aircraft failure
        - R does not consider fire as cause of system failure
    - Aircraft engine failure cuts 3 hydraulic lines, resulting in aircraft failure (DC-10 mishap)
        - R looked at triple redundant hydraulic lines
        - R looked at high rely engine
        - R did not consider CCF hazard linking engine and hydraulic line placement (safety concern)
        - R value was over optimistic

7 - 5

UAV PLOA for Reliability

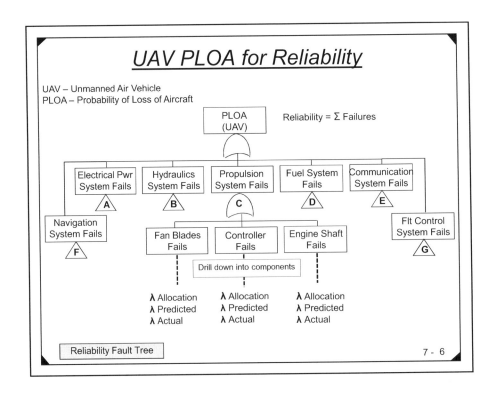

Reliability Fault Tree

7 - 6

Chapter 7 – Reliability Topics

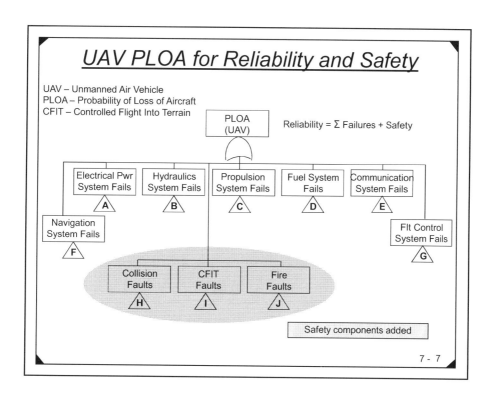

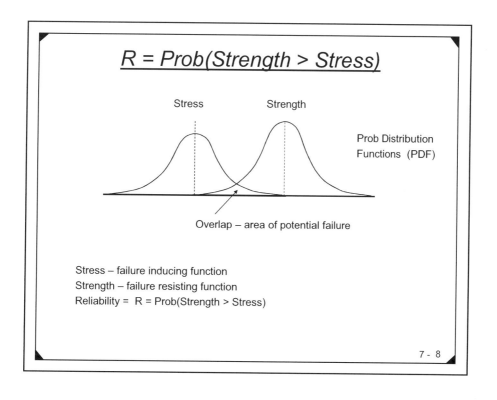

Chapter 7 – Reliability Topics

Methods for Increasing Reliability

- Classical design approach is to ensure Strength > Stress
- However, they each have some measure of variability, which is accounted for by the Reliability value

- Increase mean strength – increase size, weight, material, etc.
- Decrease average stress – control loads, use higher design dimensions
- Decrease stress variation – place limitations on use conditions
- Decrease strength variation – control process, test to eliminate less desirable parts

Design by Reliability

FRACAS

- A closed-loop corrective action process
  - Identify the *problem*
  - Investigate the root *cause* of the problem
  - Develop a *plan* which eliminates the cause
  - *Implement* the plan
  - *Verify* the effectiveness of the eliminating the cause
  - Track the entire process through problem closure

- Software is available

- Could also be used for software problem reports

- MIL-HDBK-2155, Failure Reporting, Analysis and Corrective Action Taken

Chapter 7 – Reliability Topics

Accelerated Life Testing (ALT)

- ALT is the process of testing a product by subjecting it to conditions (stress, strain, temperatures, voltage, vibration rate, pressure etc.) in excess of its normal service parameters in an effort to uncover faults and potential modes of failure in a short amount of time.
- By analyzing the product's response to such tests, its possible to make prediction about the service life and maintenance intervals of a product

Highly Accelerated Life Testing (HALT)

- HALT, is a stress testing methodology used to improve product reliability
- A technique for uncovering many of the weak links of a new product and predicting reliability
- Environmental stresses are applied in a HALT procedure, eventually reaching a level significantly beyond that expected during use
- Failure rate data used to characterize any device in a product must correlate with the stress levels in the product or application
- An environmental chamber is required for HALT testing
  - Vibration with a suitable profile in relation to frequency
  - Temperature cycling
  - Repetitive shock
  - Power margining and power cycling

Chapter 7 – Reliability Topics

Data Collection

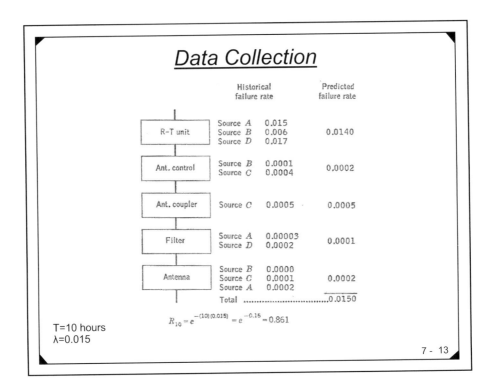

T=10 hours
λ=0.015

$R_{10} = e^{-(10)(0.015)} = e^{-0.15} = 0.861$

Derating

- Components are designed to withstand certain nominal stresses
  - They are "rated" for certain voltage, power level, frequency, etc.
- Components are rated at typical nominal stress values
- If stress increases, then FR increases
- If stress decreases, then FR decreases
- If system design is such that stress levels are intentionally lower than nominal, then the component FR can be "derated" or reduced to a lower level
- This helps reduce the overall FR of the black box

Derating is a powerful reliability design method

Chapter 7 – Reliability Topics

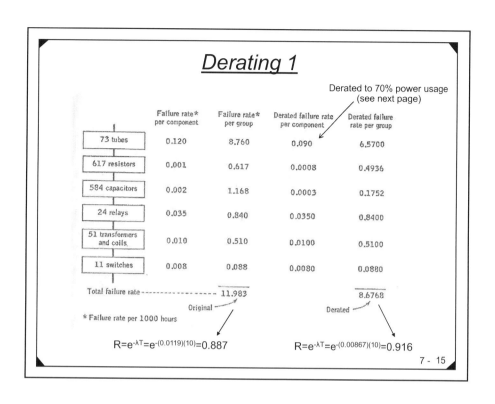

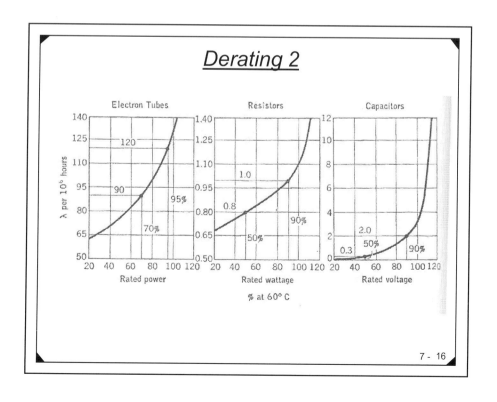

Chapter 7 – Reliability Topics

Logistics

- The management of supplies distribution and/or repair distribution
  - Supply chain
  - Repair depot
- Reliability is a major factor in logistics design
  - Inventory decisions (number of spares)
  - Repair center decisions (spares, time to repair)
  - Availability decisions (spares, operational time)
  - Sales (warranty)
- Integrated Logistics Support (ILS) is a design organization
  - Sometimes the Reliability group is under this organization

Typical Logistics Supply Channel

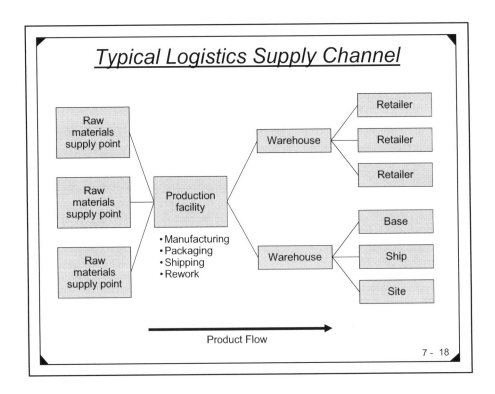

Product Flow

Chapter 7 – Reliability Topics

Typical Logistics Repair Channel

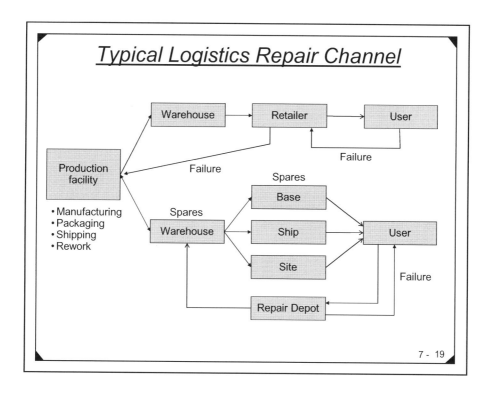

- Manufacturing
- Packaging
- Shipping
- Rework

Summary

- Quality, reliability and safety are **NOT** achieved by mathematics and statistics
  - These are only metrics for decision making
  - Quality, reliability and safety are achieved via design and manufacturing
- R is the resistance to failure of an item over time
- S is the resistance to mishaps over time
- R decrease as the number of components increases
- System R is always <= to lowest subsystem R
- Parallel (redundancy) methods increase R
- Reliability and Safety are synergistic
  - They work well together (most of the time)
  - Reliability improves Safety
  - Safety improves Reliability
- Examples
  - Two engine aircraft provides higher Safety and Reliability than a single engine
  - High reliability components help prevent safety related failures

Chapter 7 – Reliability Topics

System Safety and Reliability Analysis
Course Notes

Chapter 8
Hazard-Mishap Theory

Clifton A. Ericson II
Design Safety Solutions LLC
cliftonericson@verizon.net
540-786-3777

© C. A. Ericson II 2014

Introduction

- There is an important link between hazards, mishaps and risk
- A mishap cannot occur unless a hazard first exists
  - A mishap is the result of an actuated hazard
- Controlling hazards is critical to developing safe systems
  - By controlling hazards mishaps are prevented
  - Involves hazard identification, assessment and mitigation
- Hazards recognition requires a technical definition for a hazard
- Understanding hazard theory is critical to system safety
- Learning from accident theory is a starting point
  - An accident can be described as a scenario of "things" that happened
  - An accident scenario can result from one or more causal factors
  - Time, from accident start to finish depends upon conditions and causal factors
  - Accident causal factors can be written as a hazard scenario that predicts the accident

Chapter 8 – Hazard-Mishap Theory

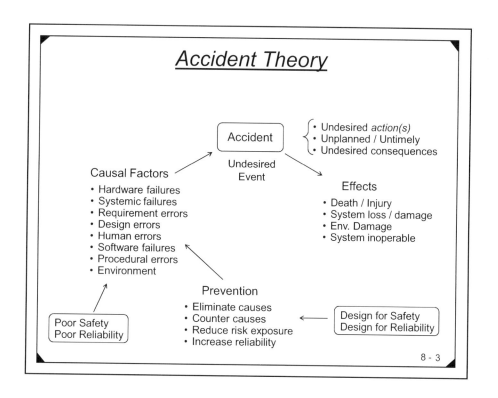

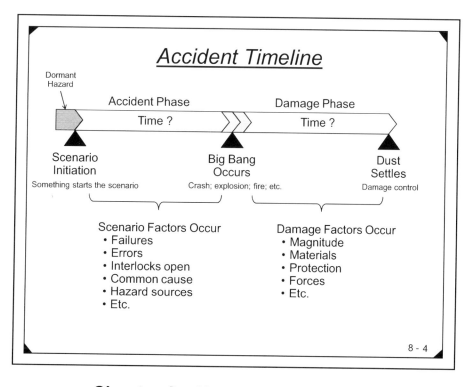

Chapter 8 – Hazard-Mishap Theory

Accident Description

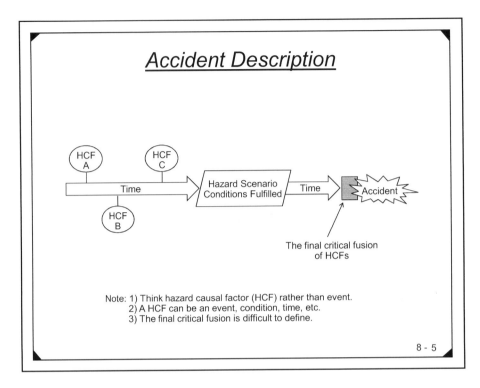

Note: 1) Think hazard causal factor (HCF) rather than event.
2) A HCF can be an event, condition, time, etc.
3) The final critical fusion is difficult to define.

8 - 5

Accident Investigation Models

1) Single event
2) Chain of Events (dominoes; multiple event chain)
3) Single Determinant Statistical Variable
4) Logic Tree (FTA of varied causal factors; multiple event dispersed)
5) Multilinear Events Sequence (breaking down into sequence of events)

After the Fact

8 - 6

Chapter 8 – Hazard-Mishap Theory

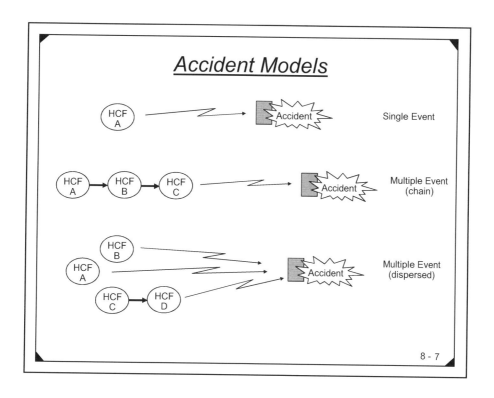

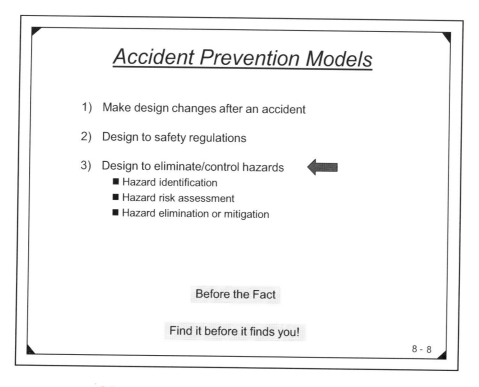

Chapter 8 – Hazard-Mishap Theory

What Is A Hazard?

- Wheel falls off car
- Car brakes fail
- Bicyclist fails to wear helmet
- Electrical wire breaks
- Hole in sidewalk
- Broken ladder rung
- Safe and Arm device fails armed
- Cracked bolt
- Broken glass
- Connector pin short
- Bomb on aircraft
- Round premature

What is a hazard?

Are these hazards?

Give some examples?

Hazard Confusion

- There is much confusion over what actually constitutes a hazard
- Intuitive, yet deceptive concept
- Industry and society definitions are vague
- Too many diverse and inconsistent hazard definitions exist
- Existing hazard definitions do not adequately describe a hazard, particularly for engineering purposes
- The vagueness of current hazards definitions leaves hazard identification open to interpretation and misunderstanding

- Need technical engineering definition
    - Specific
    - Address needs for hazard identification and mitigation

Hazard ambiguity exists and is problematic

Chapter 8 – Hazard-Mishap Theory

Industry Definitions

- A source of danger. (Dictionary)
- A condition resulting from failures, external events, errors, or combinations thereof where safety is affected. (SAE ARP-4754A)
- The presence of a potential risk situation caused by on unsafe act or condition. A condition or changing set of circumstances that presents a potential for adverse or harmful consequences; or the inherent characteristics of any activity, condition or circumstance which can produce adverse or harmful consequences. (NASA Handbook)
- A hazard is a condition or changing set of circumstances which present an injury potential. (Lawyers Desk Reference)
- A hazard is a situation that poses a level of threat to life, health, property, or environment. Most hazards are dormant or potential, with only a theoretical risk of harm; however, once a hazard becomes "active", it can create an emergency situation. A hazard does not exist when it is not happening. (Wikipedia)
- A condition, event, or circumstance that could lead to or contribute to an unplanned or undesired event. Anything, real or potential, that could make possible, or contribute to making possible, an accident. A condition that is prerequisite to an accident. (FAA)

Real Life Examples

- Signal MG71 occurs
- Oil spill
- Brake failure
- Cracked bolt
- Inadvertent missile ignition
- Controlled flight into terrain
- Explosive uncontainment of gas charge used for emergency gear extension purposes
- Explosive decompression hazard servicing tires
- Safe and arm device fails
- Electrocution
- Fire
- Fuel Leak

Not hazards; they are components of hazards

Chapter 8 – Hazard-Mishap Theory

Hazard Definition Problem

- The purpose for hazard identification is to:
    - Identify hazard potentials in the system design
    - Determine hazard causal factors
    - Determine risk
    - Determine where to improve design

- Vague definitions do not meet this criteria
    - They do not ensure the appropriate information is provided

It is critical to understand the pathology of a hazard (cause-effect)

Hazard Reality

- Current definitions do not provide guidance for identifying hazards or hazard risk
    - Certain information is needed to affect system design

- There are essentially 3 levels of hazard understanding
    - Common generic hazard
    - Pseudo hazard
    - Detailed technical hazard

Chapter 8 – Hazard-Mishap Theory

Why Hazards Exist

- Hazards exist for many different reasons
  - Component failures
  - Human error
  - Design flaws
  - The use of hazardous assets (gasoline, explosives, electricity)
  - The need for safety-critical functions (flight control)
  - Component wear, ageing, poor maintenance, etc.
  - Requirements errors
  - Combinations of above

Hazard Control

- Hazards are typically inadvertently designed into a system

- Many hazards cannot be avoided because of the system design requirements
  - Need for hazardous materials
  - Need for hazardous assets
  - Need for energy sources
  - Need for safety-critical functions

- Hazards can be prevented by countering (via design) the likelihood and potential threat
  - When hazards are not countered they present greater risk

Chapter 8 – Hazard-Mishap Theory

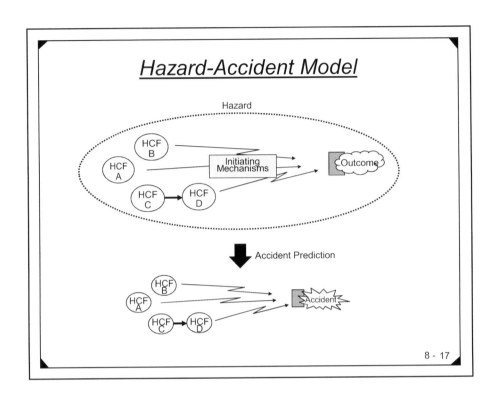

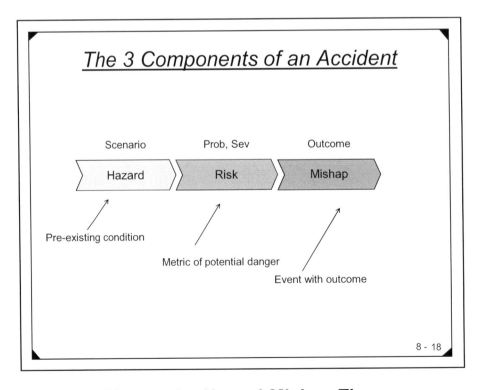

Chapter 8 – Hazard-Mishap Theory

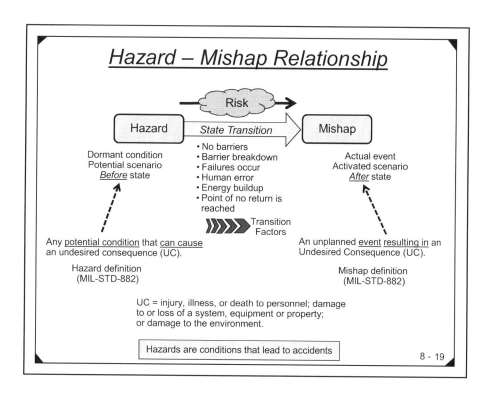

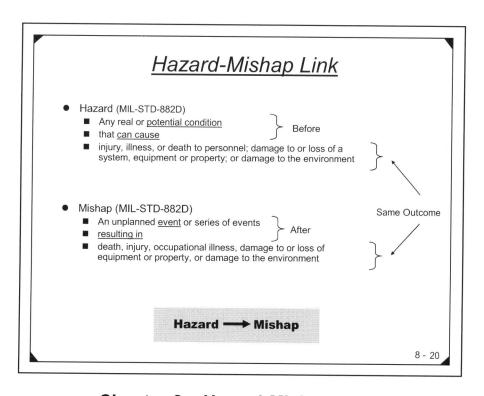

Chapter 8 – Hazard-Mishap Theory

Hazard – Mishap Relationship

```
   Hazard  ──State  Transition──▶  Mishap
 "Before" State      Elements       "After" State
Potential Condition  - Hazard Components   Actual Event (and Outcome)
                     - Risk Factors
```

- Hazards result in mishaps (I.e., hazards produce mishaps)
- A hazard is a condition that defines a possible future mishap event
- A hazard and a mishap are two different states of the same phenomenon (before and after)
- Each hazard/mishap has its own inherent and unique risk
- The probability of a hazard existing is either 1 or 0; however, the probability of a mishap is a function of the specific hazard causal factors (CFs)

8 - 21

State Transition

- There is a state transition between a Hazard and a Mishap
- Each hazard is unique, therefore the transition factors are unique to every hazard

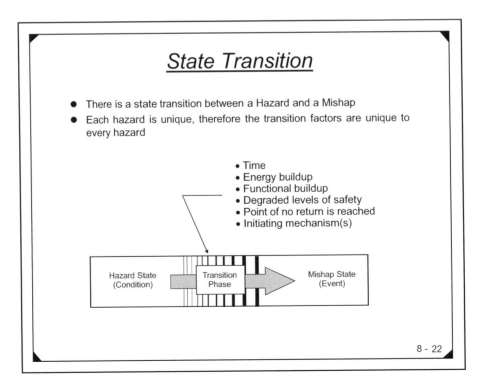

- Time
- Energy buildup
- Functional buildup
- Degraded levels of safety
- Point of no return is reached
- Initiating mechanism(s)

Hazard State (Condition) → Transition Phase → Mishap State (Event)

8 - 22

Chapter 8 – Hazard-Mishap Theory

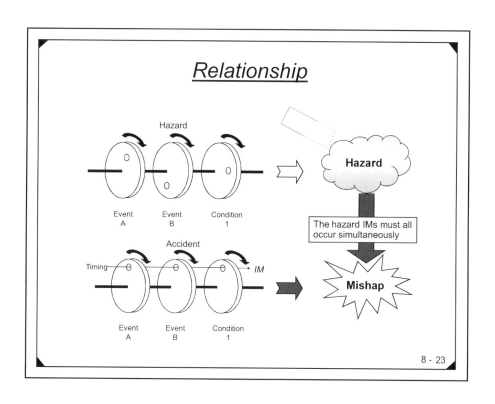

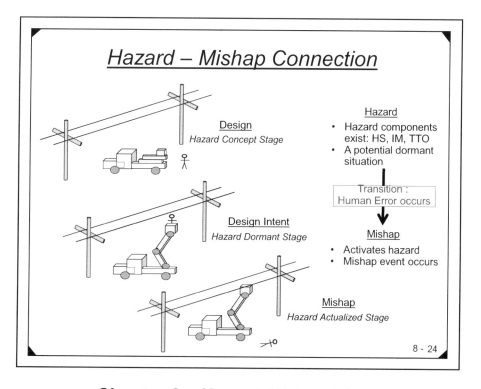

Chapter 8 – Hazard-Mishap Theory

Hazard Models

- Whatever seems dangerous (example: fuel, electricity)

- Hazard Chain – – Initiating hazard, Contributing hazard, Primary hazard [1]

- S-M-O model – – Source-Mechanism-Outcome [2]

1 – FAA System Safety Manual
2 – Pat Clemens

8 - 25

S-M-O Hazard Model
aka HS-IM-TTO Hazard Model

- Hazardous Source (HS)
  - Hazardous element
  - The basic hazardous resource creating the impetus for the hazard, such as a hazardous energy source like explosives
  - This component is usually hardware, but can be a function, environment, etc.

- Initiating Mechanism (IM)
  - Event trigger mechanism
  - This is the mechanism(s) that causes actualization of the hazard from a dormant state to an active mishap, such as a HW failure mode
  - This component can be HW, SW, human error, environment, procedures

- Target / Threat Outcome (TTO)
  - Mishap Outcome (O)
  - The person or thing that is vulnerable to injury and/or damage, such as aircraft loss due to crash
  - This component can be people, systems, environment, animals

> A hazard contains 3 basic components that are <u>necessary</u> and <u>sufficient</u> to result in a specific mishap

8 - 26

Chapter 8 – Hazard-Mishap Theory

Hazard Components

Hazard Component	Function or Purpose	Examples
Hazard Source (HS)	This is the element that provides the basic source of danger. Without it there would likely be no hazard.	• Energy sources • Safety-critical functions • Adverse environments
Initiating Mechanism(s) (IM)	These are the initiators or mechanisms that cause the mishap event to occur.	• Hardware failures • Human errors • Bent connector pins
Target-Threat Outcome (TTO)	This describes the potential outcome when the hazard becomes an actual mishap. There has to be a potential target and a threat to that target in the form of consequence and severity.	• Death or injury threat to humans • Damage threat to system or product • Damage threat to the environment

Examples

Hazardous Element	Initiating Mechanism	Target Threat Outcome
• Ordnance	• Inadvertent signal; RF energy	• Personnel / Explosion; Death/Injury
• High pressure tank	• Tank rupture	• Personnel / Explosion; Death/Injury
• Fuel	• Fuel leak and ignition source	• Personnel; System / Fire; Loss of system; Death/Injury
• High voltage	• Touching an exposed contact	• Personnel / Electrocution; Death/Injury

Chapter 8 – Hazard-Mishap Theory

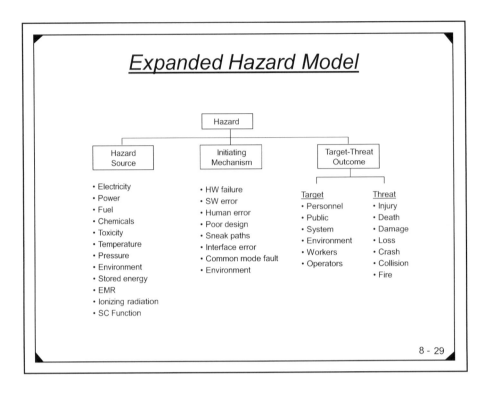

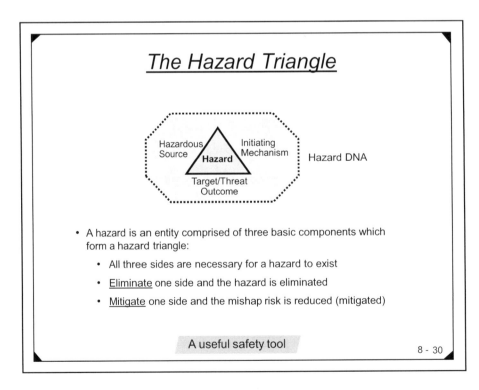

Chapter 8 – Hazard-Mishap Theory

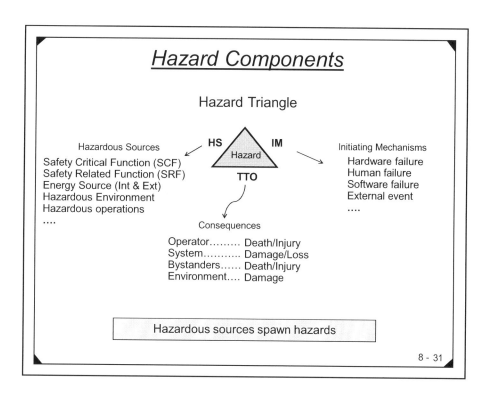

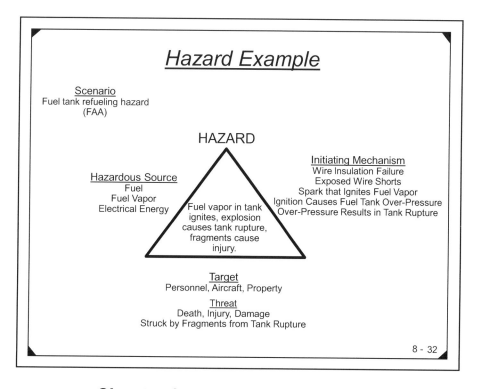

Chapter 8 – Hazard-Mishap Theory

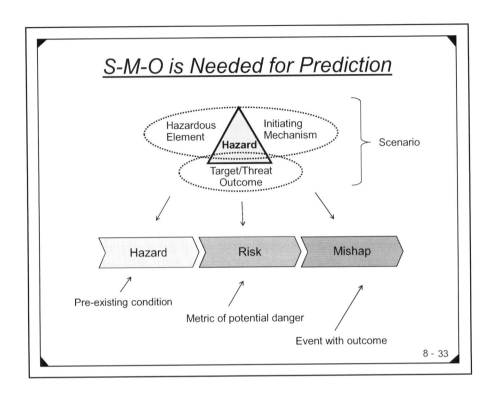

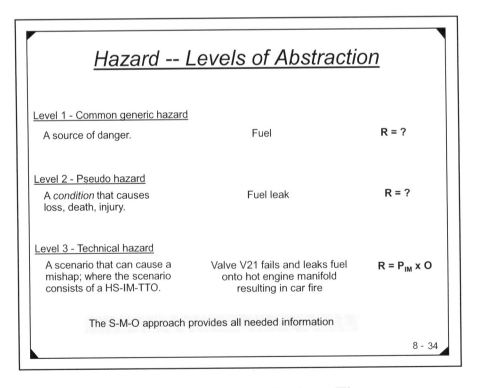

Chapter 8 – Hazard-Mishap Theory

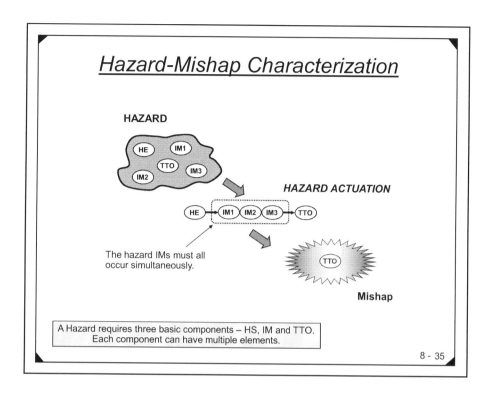

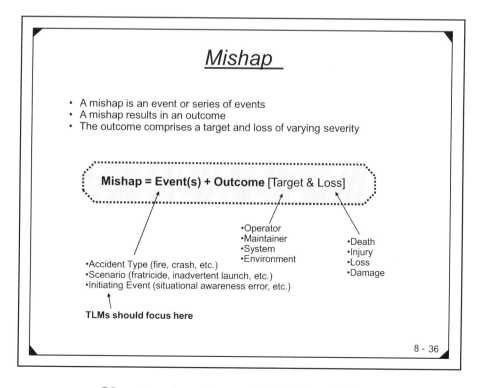

Chapter 8 – Hazard-Mishap Theory

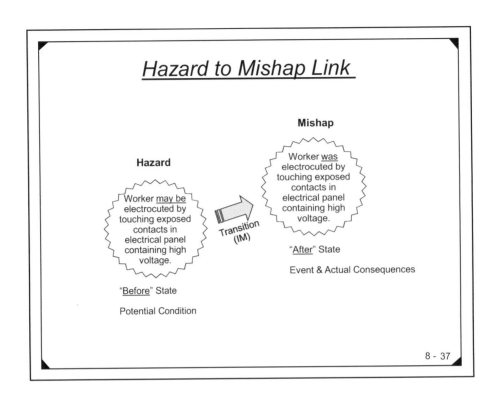

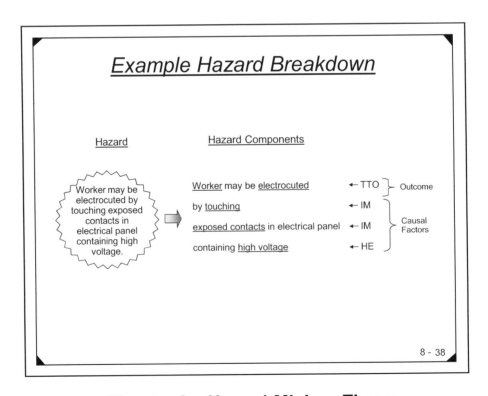

Chapter 8 – Hazard-Mishap Theory

Hazard Examples

Hazard	Source	Mechanism	Outcome
Inadvertent missile launch occurs, resulting in death or injury.	Missile with warhead	Faults / errors cause inadv launch missile	Death / injury at strike point
Automobile fuel tank ignites during collision, resulting in fire and death or injury.	Automobile fuel	Electrical spark from battery during collision	Fire, death / injury
Toaster overheats resulting in house fire.	Heating elements; voltage	Faults prevent thermal shutoff	Fire; house destroyed; Death/injury

FAA Definition Example

Note the three types of hazards, Initiating, Contributory and Primary

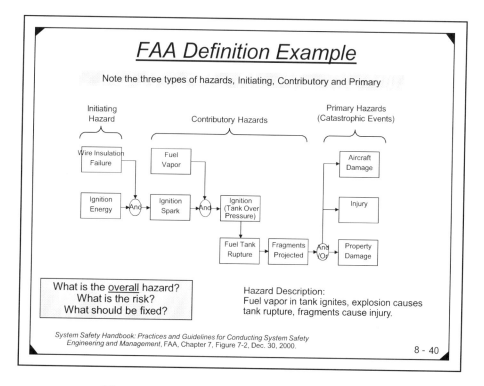

What is the <u>overall</u> hazard?
What is the risk?
What should be fixed?

Hazard Description:
Fuel vapor in tank ignites, explosion causes tank rupture, fragments cause injury.

System Safety Handbook: Practices and Guidelines for Conducting System Safety Engineering and Management, FAA, Chapter 7, Figure 7-2, Dec. 30, 2000.

Chapter 8 – Hazard-Mishap Theory

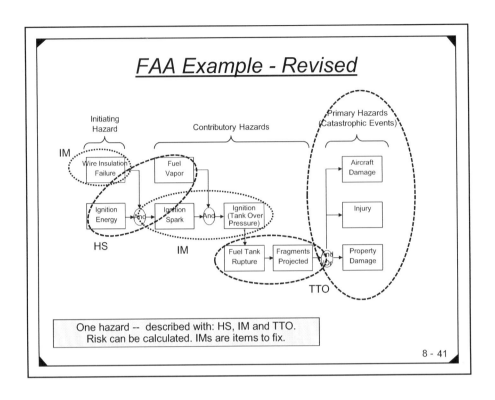

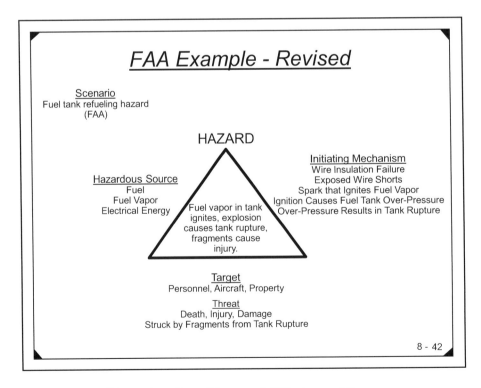

Chapter 8 – Hazard-Mishap Theory

Hazard Guidelines

- Hazard risk cannot be determined if you do not understand:
  - Scenario
  - Context
  - Specific causal factors
  - HS, IM and O provide this information
- Hazards do not cascade (i.e., Haz 1 causes Haz 2)
  - Hazard causal factors cascade
- Hazard risk cannot be properly judged when causal factors are combined
  - If the hazard probability calculation has an OR gate, then it should be two hazards

Hazard vs. Pseudo Hazard

- Pseudo Hazard
  - Part of a hazard
  - Typically the hazard outcome is states
  - Often used as a TLM
  - Example: electrocution

- Hazard
  - Full hazard description; scenario context
  - Description provides HS, IM and TTO
  - Example: failure of boom operating mechanism raises boom into high voltage power lines, resulting in worker contact with high voltage and electrocution

> Hazards should be specific and complete

Chapter 8 – Hazard-Mishap Theory

Hazard Examples – Good/Bad

Poor Examples	Good Examples
Broken glass	Worker accidentally breaks glass window and severely cuts himself from the broken glass.
High voltage	Worker is electrocuted by touching exposed contacts in electrical panel containing high voltage.
Gasoline	Automobile is hit from the rear by another auto, causing the fuel system to rupture; spilled fuel is ignited, resulting in fire that severely injures occupants.
Repair technician slips on oil.	Overhead valve V21 leaks oil on walkway below, spill is not cleaned, repair technician walking in area slips on oil and falls on concrete floor, causing serious injury.
Signal MG71 occurs.	Missile Launch signal MG71 is inadvertently generated during standby alert, causing inadvertent launch of missile and death/injury to personnel in the area of impacting missile.
Round premature	Artillery round fired from gun explodes or detonates prior to safe separation distance, resulting in death or injury to personnel within safe distance area.
Ship causes oil spill	Ship operator allows ship to run aground, causing catastrophic hull damage, causing massive oil leakage, resulting in major environmental damage.

It's important to write a complete and accurate hazard description. Include all 3 hazard components in the description.

8 - 45

Table Saw Example

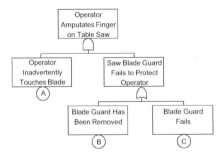

Hazard-1: "Operator amputates finger while using table saw because the operator inadvertently touches the saw blade and the blade guard is removed".

Hazard-2: "Operator amputates finger while using table saw because the operator inadvertently touches the saw blade and the blade guard fails".

8 - 46

Chapter 8 – Hazard-Mishap Theory

Definition

- A good <u>technical</u> hazard definition is needed for engineering purposes

- The HS, IM-TTO model fits this need (S-M-O)

- It establishes
  - The overall conditions involved (scenario)
  - Causal factors
  - Outcome
  - Risk

- It eliminates confusion and provides consistency

Alternate Definitions

- Hazard
  - A hazard is a precondition to a mishap; it is an existing set of specific conditions forming the potential for a mishap, that when activated becomes the actual mishap event. The conditions comprising a hazard consist of three required elements: a Hazard Source (HS), an Initiating Mechanism (IM) and a Target-Threat Outcome (TTO).

- Mishap
  - A mishap is an unplanned event resulting in undesired consequences (i.e., harm); the result of an actuated hazard. Where harm can involve death, injury, occupational illness, damage to or loss of equipment or property or damage to the environment.

Chapter 8 – Hazard-Mishap Theory

Hazard Importance

- Most system hazards cannot be eliminated
- It is important to correctly understand what a hazard is in order to identify and mitigate
- A hazard is:
    - A potential condition
    - Described as a scenario that must include:
        - Hazard Source (HS)
        - Initiating Mechanism (IM)
        - Target Threat Outcome (TTO)
    - Full hazard context is necessary (clear, concise, complete)

Hazard Guidelines

- Hazard risk cannot be determined if you do not understand:
    - Scenario
    - Context
    - Specific causal factors
    - HS, IM and TTO provide this information
- Hazards do not cascade (i.e., Haz 1 causes Haz 2)
    - Hazard *causal factors* can cascade
- Hazard risk cannot be properly judged when causal factors are combined
    - If the hazard probability calculation has an OR gate, then it should be two hazards

Chapter 8 – Hazard-Mishap Theory

Hazard vs. Pseudo Hazard

- Pseudo Hazard
  - Part of a hazard
  - Typically the hazard outcome is states
  - Often used as a TLM
  - Example: electrocution

- Hazard
  - Full hazard description; scenario context
  - Description provides HS, IM and TTO
  - Example: failure of boom operating mechanism raises boom into high voltage power lines, resulting in worker contact with high voltage and electrocution

Hazards should be specific and complete

Importance of Hazard Definition

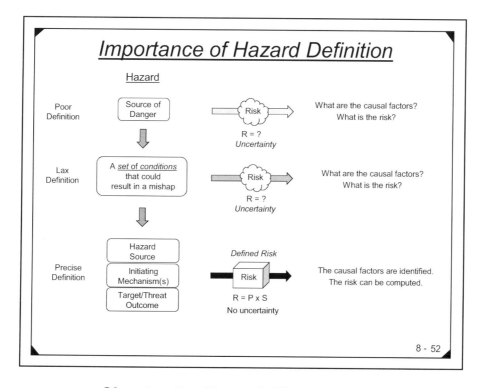

Chapter 8 – Hazard-Mishap Theory

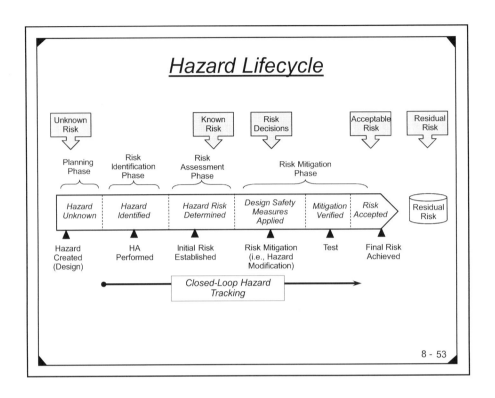

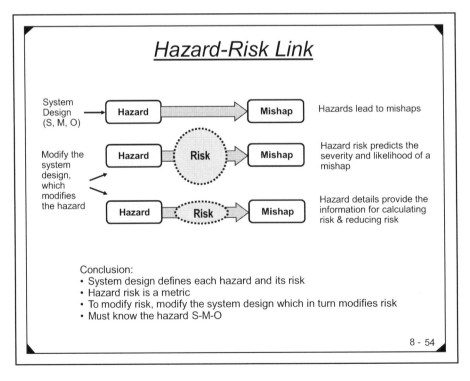

Chapter 8 – Hazard-Mishap Theory

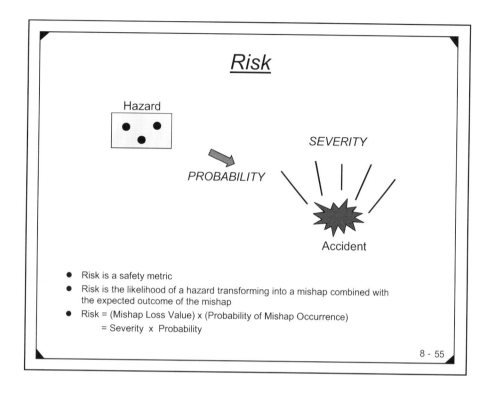

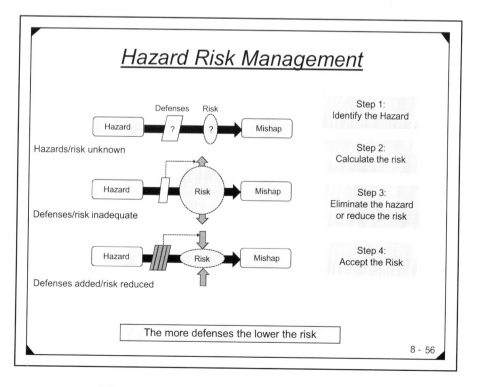

Chapter 8 – Hazard-Mishap Theory

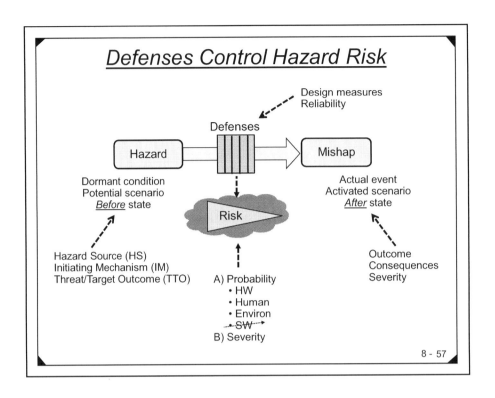

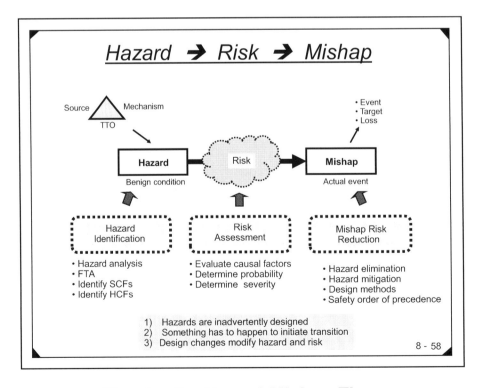

Chapter 8 – Hazard-Mishap Theory

Chapter 9 – Risk Management

What is Risk?

- Risk is a metric
- Risk is the value of a future potential event
- Risk provides decision making information

- Risk is the combination of likelihood and severity of a potential event

- Risk is a perception, based on:
    - Detailed knowledge
    - Partial knowledge
    - Past experience
    - Gut feeling
    - Guess

Which perception is best?

Types of Risk

- Cost
- Schedule
- Investment
- Product quality
- Economic
- Relational
- ………..
- ………..
- Mishap
- Hazard

Risk Options
- Prevent
- Transfer
- Accept
- Eliminate ⎫
- Reduce   ⎬ Safety Choice

There are many different types of risk

Risk Is A Reality

- We live with risk every day

- Risk will always exist

- Risk exists regardless of our knowledge, action or inaction

- We constantly make risk acceptance choices (not making a choice is making a choice)

- Risk acceptance should be based on <u>knowledge</u> not ignorance or by default

Chapter 9 – Risk Management

How is Safety Measured?

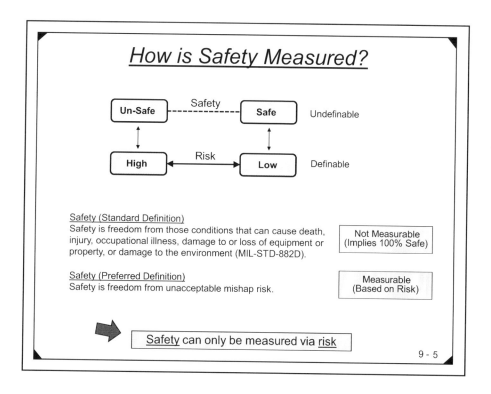

Safety (Standard Definition)
Safety is freedom from those conditions that can cause death, injury, occupational illness, damage to or loss of equipment or property, or damage to the environment (MIL-STD-882D).

Not Measurable (Implies 100% Safe)

Safety (Preferred Definition)
Safety is freedom from unacceptable mishap risk.

Measurable (Based on Risk)

➡ Safety can only be measured via risk

9 - 5

Safety Risk

- Risk is a safety metric

- Risk is an expression of the impact and possibility of a mishap in terms of potential mishap severity and probability of occurrence (MIL-STD-882D)

- Risk = (Mishap Loss Value) x (Probability of Mishap Occurrence)
  = Severity x Probability

- Risk is a criticality measure of hazard significance
  - Qualitative
  - Quantitative

Risk provides the measure of a hazard

9 - 6

Chapter 9 – Risk Management

Risk

- Risk is a safety metric that is associated with hazards
- Risk predicts the likelihood of a mishap and the potential outcome
- Risk is foreseeable and preventable
- Risk can be controlled/managed
- Potential mishaps are prevented via the <u>risk management</u> aspect of <u>system safety</u>

- The following terms are essentially equivalent:
  - Risk
  - Hazard risk
  - Mishap risk
  - Safety risk

Hazard – Risk – Mishap

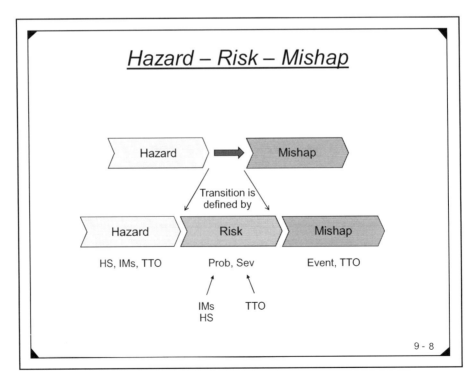

Chapter 9 – Risk Management

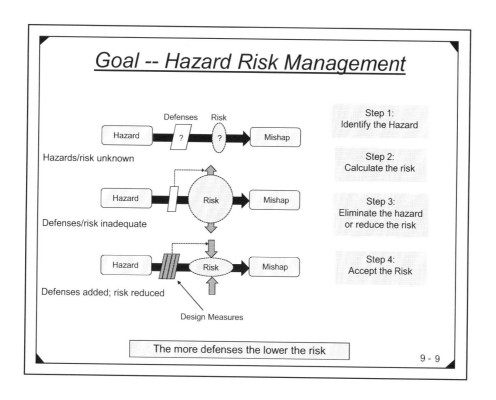

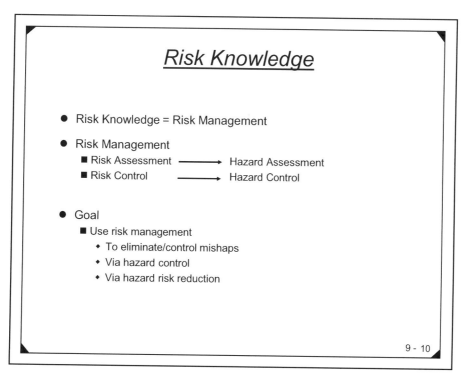

Chapter 9 – Risk Management

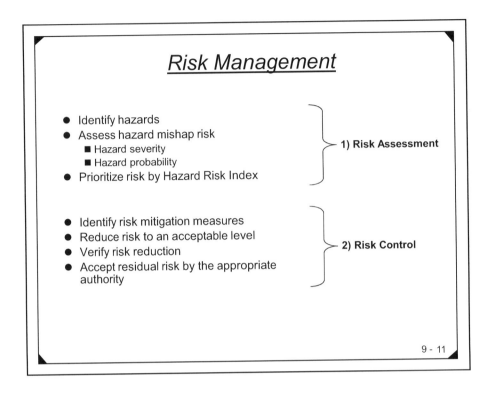

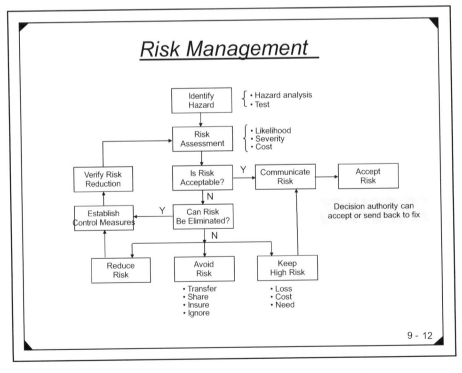

Chapter 9 – Risk Management

Risk Management Considerations

- Risk Assessment
  - Determine level of risk
  - Must include units
  - Confidence level of data/results
- Risk Acceptance
  - Judgment criteria
  - Whose value Is it Safe for (Utility function)
- Risk Communication
  - Purpose
  - Timely, appropriate, correct
  - Trust in messenger (politicians less, academics more)
  - Context/meaning (clear, misleading)

Risk Factors

- Risk involves a future event characterized by Probability and Severity

- Probability
  - Probability that the hazard transitions to a mishap
  - Use expected worst case
  - P = function(HS, IMs, TTO)

- Severity
  - Severity is the expected mishap outcome loss
  - Use expected worst case
  - S = function(HS, TTO)

Chapter 9 – Risk Management

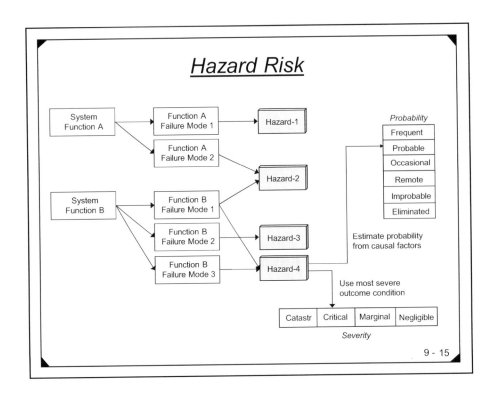

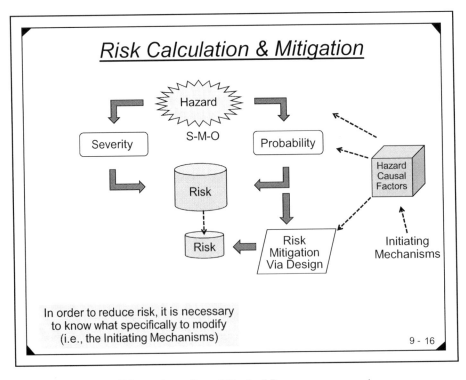

Chapter 9 – Risk Management

Risk Calculations

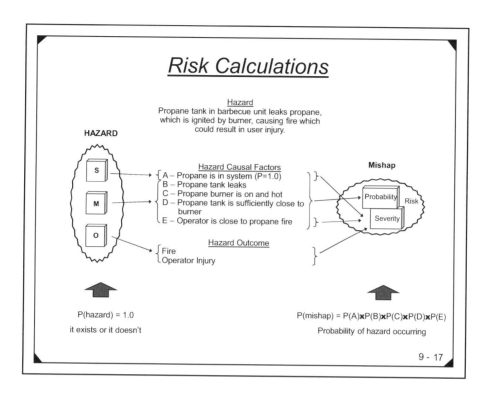

Hazard
Propane tank in barbecue unit leaks propane, which is ignited by burner, causing fire which could result in user injury.

Hazard Causal Factors
- A – Propane is in system (P=1.0)
- B – Propane tank leaks
- C – Propane burner is on and hot
- D – Propane tank is sufficiently close to burner
- E – Operator is close to propane fire

Hazard Outcome
- Fire
- Operator Injury

P(hazard) = 1.0
it exists or it doesn't

P(mishap) = P(A)×P(B)×P(C)×P(D)×P(E)
Probability of hazard occurring

9 - 17

Risk Evaluation Methods

- HRI Matrix
- ALARP
- Bar Chart

- Considerations
  - Qualitative or Quantitative
  - Flexibility
  - Cost
  - Risk acceptance

Select method for program

9 - 18

Chapter 9 – Risk Management

Qualitative vs. Quantitative

- Qualitative Risk Assessment
  - Mishap Risk Index (MRI) establishes semi-quantitative criteria
  - Risk Acceptance Matrix established risk decision criteria

- Quantitative Risk Assessment
  - Probabilistic Risk Assessment (PRA)
    - Fault Tree Analysis
    - Event Tree Analysis

Bar Chart

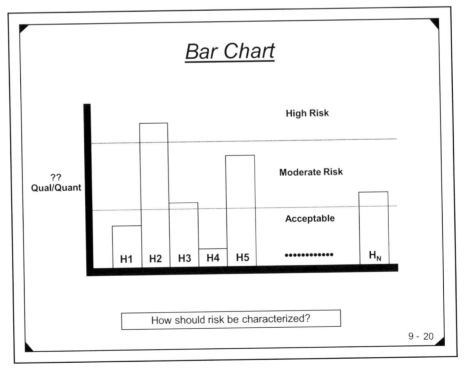

How should risk be characterized?

Chapter 9 – Risk Management

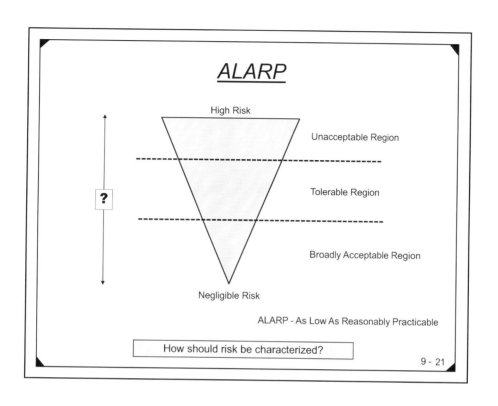
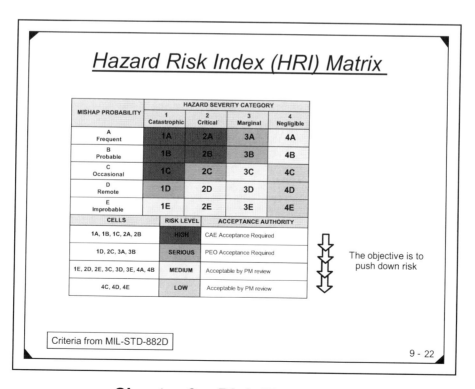

Chapter 9 – Risk Management

Hazard Risk Index (HRI) Matrix

Probability	Severity			
	1 Catastrophic	2 Critical	3 Marginal	4 Negligible
A - Frequent	1	3	7	13
B - Probable	2	5	9	16
C - Occasional	4	6	11	18
D - Remote	8	10	14	19
E - Improbable	12	15	17	20
F - Eliminated	Eliminated			

High / Serious / Medium / Low

The objective is to push down risk

Criteria from MIL-STD-882E

MIL-STD-882E Risk Measures

PROBABILITY LEVELS		SEVERITY LEVELS	
LEVEL	CRITERIA	LEVEL	CRITERIA
A Frequent	Likely to occur often in the life of an item	1 Catastrophic	Could result in one or more of the following: death, permanent total disability, irreversible significant environmental impact, or monetary loss equal to or exceeding $10M.
B Probable	Will occur several times in the life of an item.	2 Critical	Could result in one or more of the following: permanent partial disability, injuries or occupational illness that may result in hospitalization of at least three personnel, reversible significant environmental impact, or monetary loss equal to or exceeding $1M but less than $10M.
C Occasional	Likely to occur sometime in the life of an item.		
D Remote	Unlikely, but possible to occur in the life of an item.		
E Improbable	So unlikely, it can be assumed occurrence may not be experienced in the life of an item.	3 Marginal	Could result in one or more of the following: injury or occupational illness resulting in one or more lost work day(s), reversible moderate environmental impact, or monetary loss equal to or exceeding $100K but less than $1M.
F Eliminated	Incapable of occurrence. This level is used when potential hazards are identified and later eliminated.	4 Negligible	Could result in one or more of the following: injury or occupational illness not resulting in a lost work day, minimal environmental impact, or monetary loss less than $100K.

Chapter 9 – Risk Management

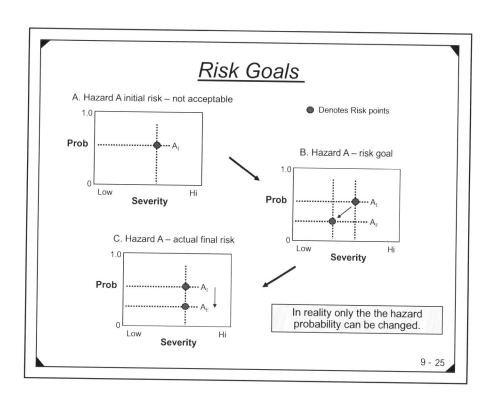

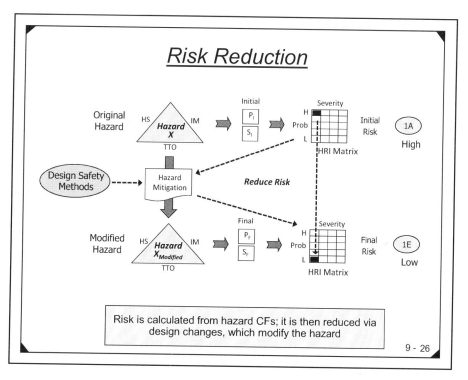

Chapter 9 – Risk Management

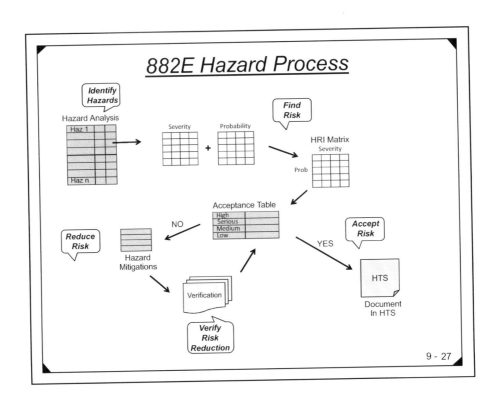

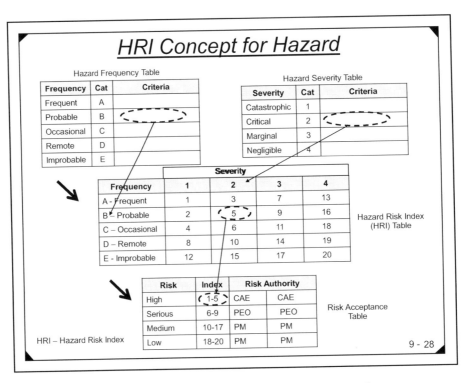

Chapter 9 – Risk Management

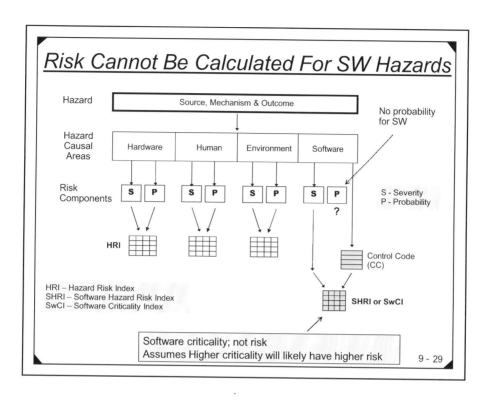

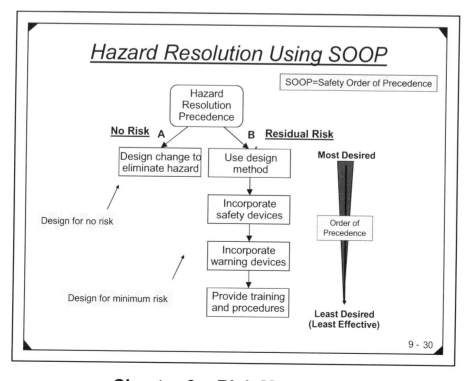

Chapter 9 – Risk Management

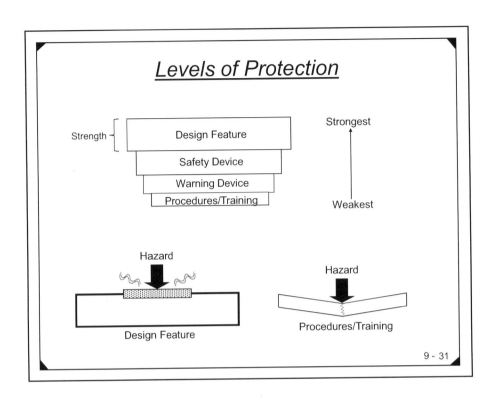

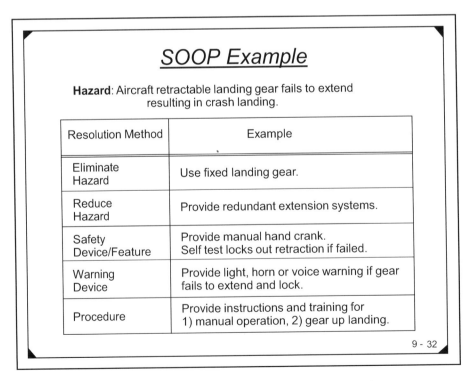

Chapter 9 – Risk Management

Risk Rules

- If a hazard is eliminated, then there is no risk
- If a hazard cannot be eliminated, then the hazard, and it's associated risk, always exist (even if mitigated)
- Hazard risk can be reduced by design methods (mitigation)
  - Always able to reduce probability
  - Seldom able to reduce severity
- System safety goals
  - Eliminate the hazard, OR
  - Reduce the risk (to a low & acceptable level)
- Residual risk
  - The remaining risk level after mitigation techniques have been applied
- Risk resulting from HW failures increases exponentially with time because $P = 1 - e^{-\lambda T}$

Hazard Risk vs. Mishap Risk

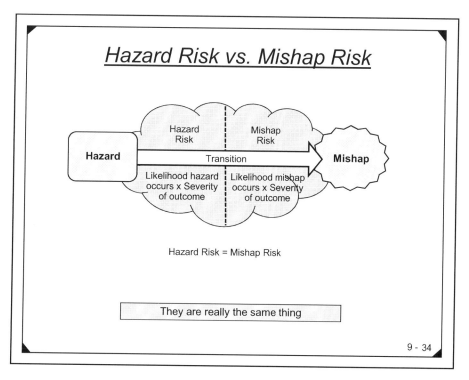

Hazard Risk = Mishap Risk

They are really the same thing

Chapter 9 – Risk Management

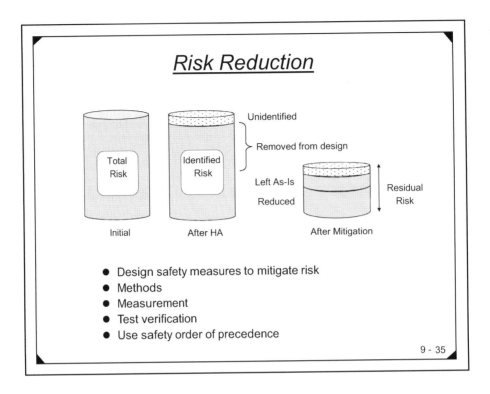

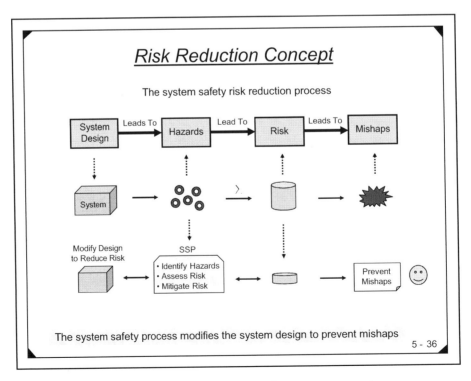

Chapter 9 – Risk Management

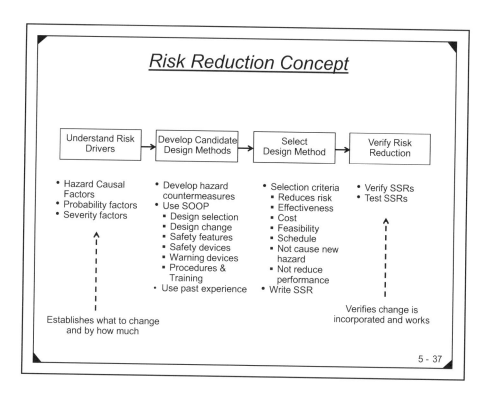

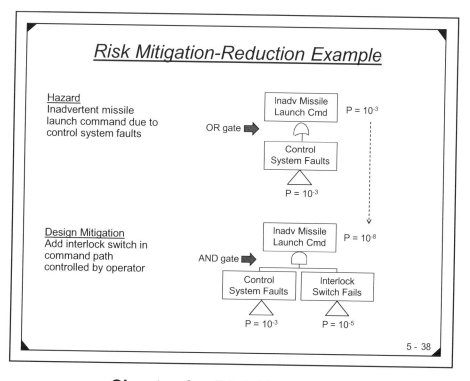

Chapter 9 – Risk Management

Layered Safety Protection

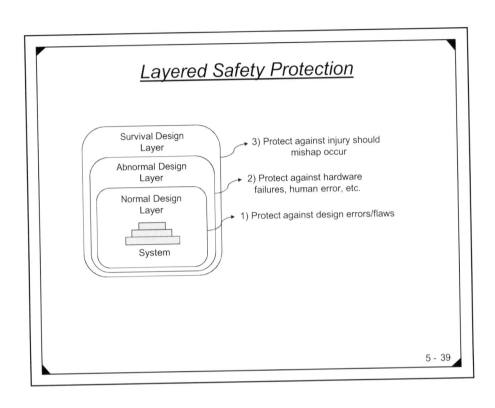

- Survival Design Layer → 3) Protect against injury should mishap occur
- Abnormal Design Layer → 2) Protect against hardware failures, human error, etc.
- Normal Design Layer → 1) Protect against design errors/flaws
- System

Design Safety Methods

	DSF	SOOP Category
1	Redundancy design	Design
2	Fail-safe design	Design
3	Fault-tolerant design	Design
4	No single-point failure design	Design
5	Design diversity	Design
6	Barriers	Design
7	Partitions	Design
8	Interlock (inhibit) design	Design
9	Intrinsic Safety design	Design
10	Isolation design	Design

Chapter 9 – Risk Management

Design Safety Methods

	DSF	SOOP Category
11	Enhanced part reliability	Design
12	Weak link design	Design
13	Safety factors	Design
14	Safety margins	Design
15	Fail-operational design	Design
16	Fail-non-operational design	Design
17	Energy control design	Design
18	Fire detection and suppression	Safety Device
19	Personal protective equipment (PPE)	Safety Device
20	Survival system design	Safety Device

Design Safety Methods

	DSF	SOOP Category
21	Lockout/Tagout	Safety Device
22	Warnings and Cautions	Warning Device
23	Special training	Training
24	Special procedures	Procedure
25	Scheduled X-ray analysis before use	Procedure
26	Scheduled maintenance before use	Procedure
27	Scheduled inspection before use	Procedure

Chapter 9 – Risk Management

Common Problems with Mitigation Methods

- Inappropriate for the problem
- Operators dislike it
- Managers dislike it
- Too costly (unsustainable)
- Overmatched by other priorities
- Misunderstood
- Progress measured too late
- Insufficient or inadequate documentation and training

Summary (1)

- Risk = Severity x Probability

- Working with qualitative risk tables is much easier and quicker than working with quantitative risk values

- Risk increases as severity and/or probability increases (and vice-versa)

- When attempting to reduce risk, risk (hazard) probability can always be changed, however, it is difficult to change risk (hazard) severity

Chapter 9 – Risk Management

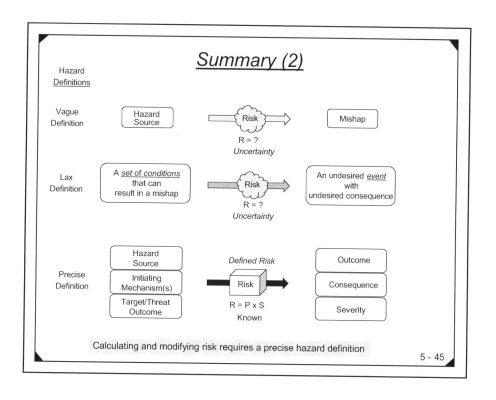

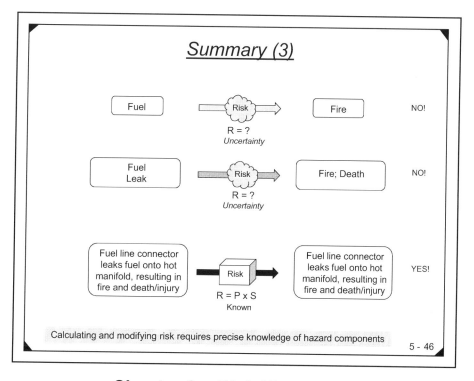

Chapter 9 – Risk Management

System Safety and Reliability Analysis
Course Notes

Chapter 10
Hazard Analysis Techniques

Clifton A. Ericson II
Design Safety Solutions LLC
cliftonericson@verizon.net
540-786-3777

© C. A. Ericson II 2014

Introduction

- Hazard analysis (HA) is the process for identifying hazards
- A HA technique is typically utilized for HA
- Many different HA techniques exist
- Certain HA techniques are invaluable for certain problems and certain design phases
- Care must be taken to use the right HA technique(s) for a particular application
- Each HA technique has a unique set of attributes

Chapter 10 – Hazard Analysis Techniques

What Is Hazard Analysis Technique?

- A methodology that assists the analyst in the discovery of hazards

- The methodology should be:
  - Rigorous
  - Systematic
  - Thorough – System to component coverage

Attributes of HA Techniques

- Primary vs. Secondary
- Qualitative vs. Quantitative
- Inductive vs. Deductive
- Type vs. Technique

- Timing (when impacts design)
- Detail (data required; data available;)
- Scope (system coverage; ground rules)
- Time required
- HA Complexity (time; expertise; difficulty)
- Cost
- Difficulty
- Expertise required
- Objectives of the analysis
- Subjectivity

Chapter 10 – Hazard Analysis Techniques

HA: Primary vs. Secondary

- Primary HA Technique
  - A complete hazard identification technique
  - E.g., PHA, SHA

- Secondary HA Technique
  - An incomplete hazard identification technique
  - Typically supports a primary technique
  - E.g., FMEA, FTA

HA: Qualitative vs. Quantitative

- Qualitative
  - Qualitative results
  - Involves the use of qualitative criterion
  - More subjective
  - Uses expert opinion

- Quantitative
  - Quantitative results
  - Involves the use of numerical or quantitative data
  - More objectivity
  - Results can be biased by the validity and accuracy of the input numbers
  - PRA - probabilistic risk assessment

Chapter 10 – Hazard Analysis Techniques

HA: Qualitative vs. Quantitative

	Attribute	Qualitative	Quantitative
1	Numerical results	No	Yes
2	Cost	Lower	Higher
3	Subjective/Objective	Subjective	Objective
4	Difficulty	Lower	Higher
5	Complexity	Lower	Higher
6	Data	Less detailed	More detailed
7	Technical Expertise	Lower	Higher
8	Time Required	Lower	Higher
9	Tools required	Seldom	Usually
10	Accuracy	Lower	Higher

Don't confuse mathematical model results with reality

HA: Inductive vs. Deductive

- Inductive reasoning
    - A conclusion is proposed that contains more information than the observation or experience on which it is based
    - An inductive hazard analysis might conclude more than the given data intends to yield
    - Example - FMEA; "if resistor R44 fails the system will shutdown"

- Deductive reasoning
    - A conclusion is drawn from a set of premises and contains no more information than the premises taken collectively
    - A deductive hazard analysis would conclude no more than the data provides (direct cause and effect)
    - Example - FTA; "inadvertent launch requires a power fault and a command signal fault"

Does it really matter to the analyst?

Chapter 10 – Hazard Analysis Techniques

HA: Inductive vs. Deductive

	Inductive	Deductive
Methodology	– What-If – Going from the specific to the general	– How-Can – Going from the general to the specific
General characteristics	– System is broken down into individual components – Potential failures for each component are considered (what can go wrong?) – Effects of each failure are defined (what happens if it goes wrong?)	– General nature of the hazard has already been identified (fire, inadvertent launch, etc.) – System is reviewed to define the cause of each hazard (how can it happen?)
Applicability	– Systems with few components – Systems where Single Point Failures (SPFs) are predominant – Preliminary or overview analysis	– All sizes of systems – Developed for complex systems – Designed to identify hazards caused by multiple failures
Potential pitfalls	– Difficult to apply to complex systems – Large number of components to consider – Consideration of failure combinations becomes difficult	– Detailed system documentation required – Large amount of data involved – Time consuming
Examples	– FMEA – HAZOP – FHA – PHA	– Fault Tree Analysis (FTA) – Event Tree Analysis (ETA) – Common Cause Failure Analysis – PHA

10 - 9

Top-Down or Bottom-Up

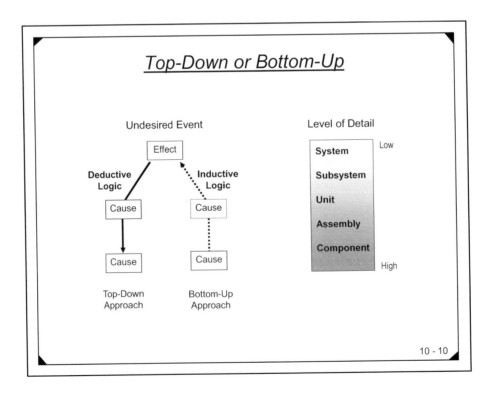

10 - 10

Chapter 10 – Hazard Analysis Techniques

HA: Type vs. Technique

- Analysis "Type"
  - Establishes a general category of analysis type
  - The category defines detail, timing, coverage (What, Where and When to analyze)
  - Establishes a specific analysis task for a specific time in program lifecycle
  - Provides a specific design focus
  - Seven basic types

- Analysis "Technique"
  - Establishes a specific and unique analysis methodology
  - Defines the principles and procedures of HA inquiry
  - Establishes how to perform the analysis
  - Satisfies the intent of a specific HA Type
  - > 100 techniques (see SSS Safety Analysis Handbook)

The 7 Prime HA Types

Preliminary Hazard List (PHL) – Identifies general system level hazards and mishaps. Identify safety guidelines, precepts, TLMs and SCFs.

Preliminary Hazard Analysis (PHA) – Identifies system hazards, causal factors and risk. Start identifying safety guidelines, SSRs, precepts, SCFs and TLMs.

Subsystem Hazard Analysis (SSHA) – Identify hazards at subsystem level and interface level. Concentrate on detailed causal factors within the subsystem.

System Hazard Analysis (SHA) – Assess the risk of the total system design, specifically of the subsystem interfaces. Include HW, SW, HSI, COTS, etc.

Operations & Support Hazard Analysis (O&SHA) – Identify hazards at operations and support level, considering human error and design.

Health Hazard Assessment (HHA) – Identify hazards to humans resulting from design, operation and manufacturing (eg, noise, ergonomics, HazMat).

Safety Requirements/Criteria Analysis (SRCA) – Ensure all hazards are covered by SSRs, and all SSRs are in design and test specifications and successfully pass testing.

Ref: MIL-STD-882

Chapter 10 – Hazard Analysis Techniques

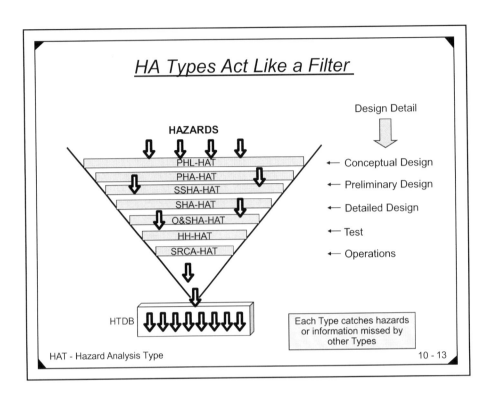

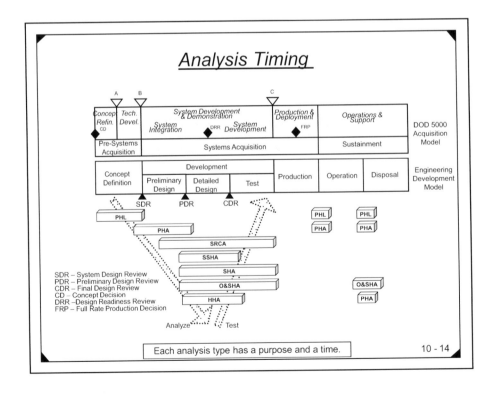

Chapter 10 – Hazard Analysis Techniques

HA Relationships

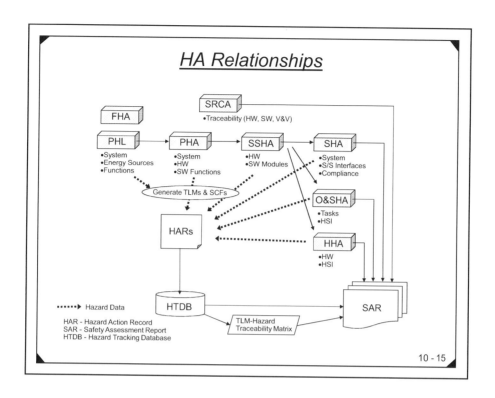

HA Techniques

- There are many proposed HA techniques
- The Safety Analysis Handbook produced by the ISSS lists over 100 HA techniques
- Use caution
    - Many techniques are derivations of other techniques and many are not very useful

- Application guidance
    - Use techniques required by contract or regulation
    - Use techniques that work best for the application

Chapter 10 – Hazard Analysis Techniques

Most Common HA Techniques

Acronym	Name	Type	Technique	Primary	Secondary
PHL	Preliminary Hazard List	X	**X**	X	
PHA	Preliminary Hazard Analysis	X	**X**	X	
SSHA	Subsystem Hazard Analysis	X	**X**	X	
SHA	System Hazard Analysis	X	**X**	X	
O&SHA	Operating & Support Hazard Analysis	X	**X**	X	
HHA	Health Hazard Assessment	X	**X**	X	
SRCA	Safety Requirements & Criteria Analysis	X	**X**		X
FTA	Fault Tree Analysis		**X**		X
ETA	Event Tree Analysis		**X**		X
FMEA	Failure Modes and Effects Analysis		**X**		X
FHA	Functional Hazard Analysis		**X**	X	
FHA	Fault Hazard Analysis		**X**	X	
HAZOP	Hazard and Operability Study		**X**	X	
SCA	Sneak Circuit Analysis		X		X
BA	Barrier Analysis		X		X
BPA	Bent Pin Analysis		X		X
THA	Threat Hazard Assessment		X		X
CCA	Cause Consequence Analysis		X		X
CCFA	Common Cause Failure Analysis		X		X
MORT	Management Oversight and Risk Tree		X		X
PNA	Petri Net Analysis		X		X
MA	Markov Analysis		X		X

10 - 17

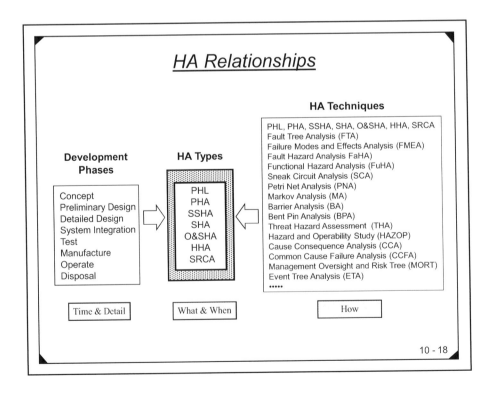

10 - 18

Chapter 10 – Hazard Analysis Techniques

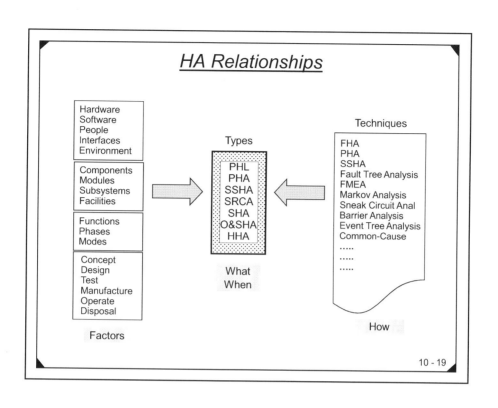

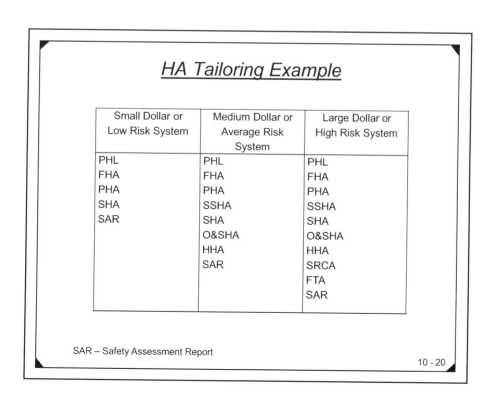

Chapter 10 – Hazard Analysis Techniques

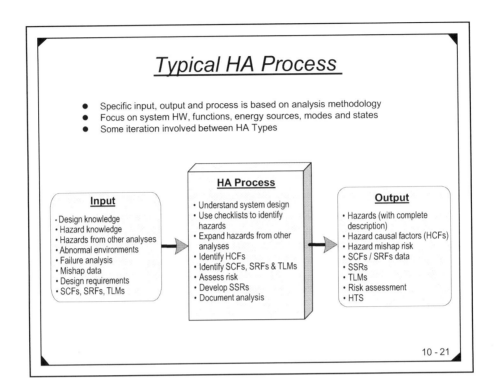

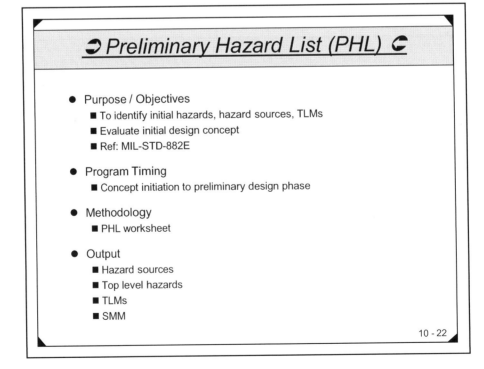

Chapter 10 – Hazard Analysis Techniques

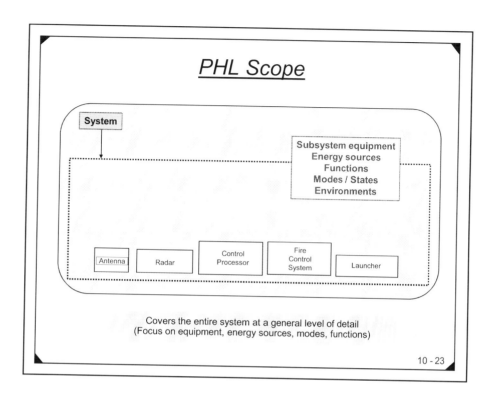

Chapter 10 – Hazard Analysis Techniques

Preliminary Hazard Analysis (PHA)

- Purpose / Objectives
  - To identify system hazards, hazard sources, TLMs
  - Evaluate preliminary design
  - Ref: MIL-STD-882E

- Program Timing
  - Preliminary design phase

- Methodology
  - PHA worksheet

- Output
  - System hazards
  - Hazard sources
  - TLMs
  - SMM

PHA Scope

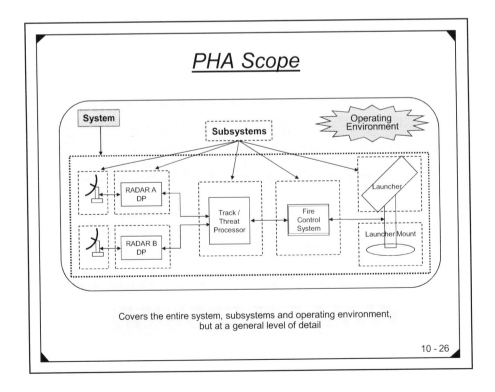

Covers the entire system, subsystems and operating environment, but at a general level of detail

Chapter 10 – Hazard Analysis Techniques

PHA Worksheet

System _____   Analyst _____
Subsystem _____   Date _____

Hazard	Cause	Effect	IHRI	Recommended Action	FHRI	Remarks	Status

Note: Worksheets can be modified

➲ Subsystem Hazard Analysis (SSHA) ⊂

- Purpose / Objectives
  - To identify detailed subsystem hazards
  - Evaluate subsystem design
  - Ref: MIL-STD-882E

- Program Timing
  - Subsystem design phase

- Methodology
  - SSHA worksheet

- Output
  - Subsystem hazards
  - Hazard sources
  - TLMs
  - SMM

Chapter 10 – Hazard Analysis Techniques

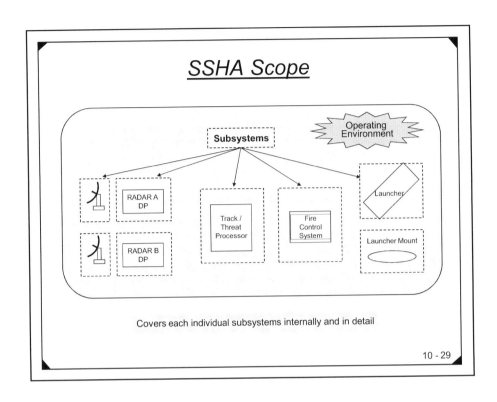

Chapter 10 – Hazard Analysis Techniques

System Hazard Analysis (SHA)

- Purpose / Objectives
  - To identify system hazards, TLMs
  - Evaluate system design
  - Ref: MIL-STD-882E
- Program Timing
  - Final design phase
- Methodology
  - SHA worksheet
- Output
  - System hazards
  - Hazard sources
  - TLMs
  - SMM

SHA Scope

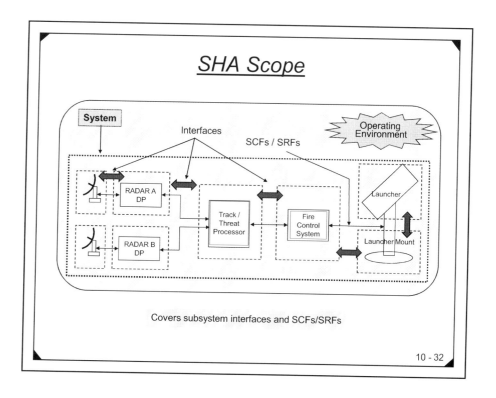

Covers subsystem interfaces and SCFs/SRFs

Chapter 10 – Hazard Analysis Techniques

SHA Worksheet

System _____   Analyst _____
Subsystem _____   Date _____

Hazard	Cause	Effect	IHRI	Recommended Action	FHRI	Remarks	Status

Note: Worksheets can be modified

Operations & Support HA (O&SHA)

- Purpose / Objectives
  - To identify operational level hazards, hazard sources, TLMs
  - Evaluate the operations concept and support equipment
  - Ref: MIL-STD-882E

- Program Timing
  - Final design phase through testing phase

- Methodology
  - O&SHA worksheet

- Output
  - Operational hazards
  - Support equipment hazards
  - SMM

Chapter 10 – Hazard Analysis Techniques

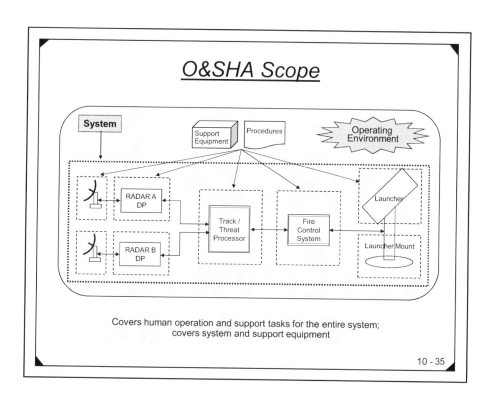

Chapter 10 – Hazard Analysis Techniques

Health Hazard Assessment (HHA)

- Purpose / Objectives
  - To identify hazards contributing to human health
  - Evaluate design for health hazards
  - Ref: MIL-STD-882E
- Program Timing
  - Preliminary and final design phases
- Methodology
  - HHA worksheet
- Output
  - Health hazards
  - Hazard sources
  - TLMs

HHA Worksheet

System_____   Analyst_____
Subsystem_____   Date_____

Human Health Factor	Hazard	Cause	Effect	IHRI	Recommended Action	FHRI	Status

Note: Worksheets can be modified

Chapter 10 – Hazard Analysis Techniques

Functional Hazard Analysis (FHA)

- Purpose / Objectives
    - To identify system functional hazards
    - Evaluate preliminary/detailed design
    - Focuses on system functions
    - Ref: MIL-STD-882E and SAE ARP 4754/4761

- Program Timing
    - Preliminary design phase

- Methodology
    - PHA worksheet

- Output
    - System hazards
    - Hazard sources
    - TLMs
    - SMM

- Functions are a critical aspect of system design.
- Safety-critical functions (SCFs) are always a source for hazards.

10 - 39

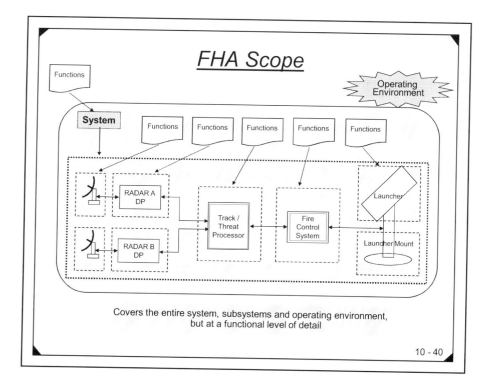

Covers the entire system, subsystems and operating environment, but at a functional level of detail

10 - 40

Chapter 10 – Hazard Analysis Techniques

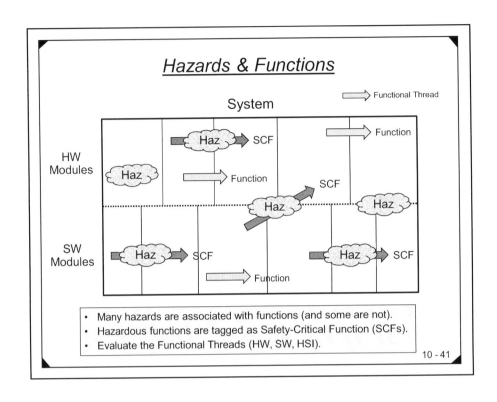

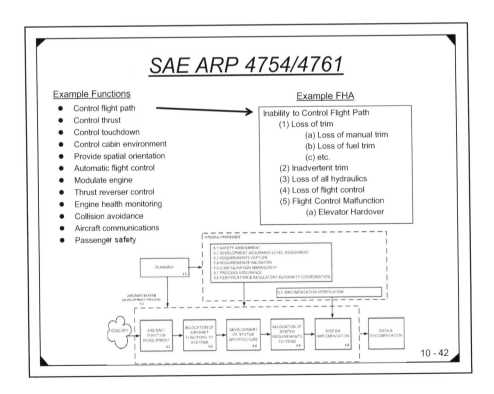

Chapter 10 – Hazard Analysis Techniques

FHA

Function	Potential Anomaly List							
	Fails	Inadvertent	Erratic	Occurs Late	Occurs Early	Slow	Input Error	Output Error
Funct 1								
Funct 2								
Funct 3								
Funct 99								

→ Build List

Function Table

- Build potential anomaly list and Function table.
- Evaluate each function for all potential anomalies.
- Hazards and SCFs will develop from the analysis.
- SCF threads identify SC HW, SW and HIS.

10 - 43

Functional Hazard Analysis

Anomaly List Considerations
- Fails to function
- Malfunction
- Degraded function
- Inadvertent function
- Incorrect input to function
- Incorrect output from function
- Function timing error – early, late, slow
- Unable to stop/control
- Operator confusion
- Out of sequence
- Erratic function

System Considerations
- Performance
- Control
- Display
- Detection
- Operator
- Computation
- Data

10 - 44

Chapter 10 – Hazard Analysis Techniques

FHA Worksheet

System _____   Analyst _____
Subsystem _____   Date _____

Function	Potential Anomaly	Hazard	IHRI	Recommended Action	FHRI	Remarks	Status

Note: Worksheets can be modified

Summary

- Hazard analysis (HA) is the process for identifying hazards
- A HA technique is typically utilized for HA
- HA worksheets can be tailored
- Care must be taken to use the right HA technique(s) for a particular application
- HA is more complex than realized
- HA requires an experienced analyst (or an analyst being mentored)
- A HA should be peer reviewed for credibility and accuracy

Chapter 10 – Hazard Analysis Techniques

Chapter 11 – Hazard Recognition

Introduction

- There are four critical aspects to hazard analysis (HA)
  - Knowing what comprises a hazard
  - Knowing how to identify (recognize) hazards
  - Knowing how to correctly describe a hazard
  - Knowing how to organize hazards

- This chapter focuses on hazard recognition

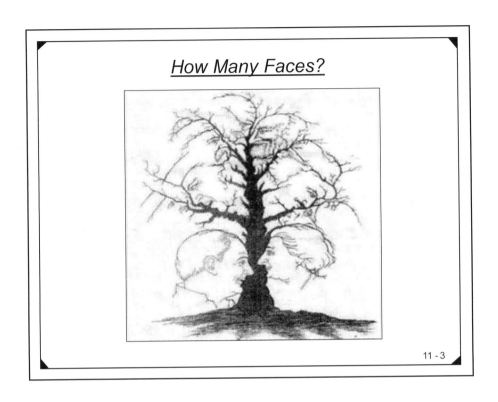

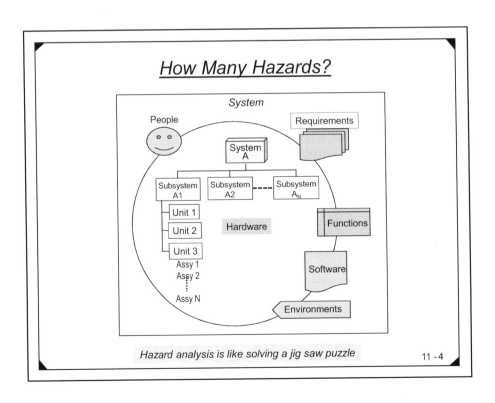

Chapter 11 – Hazard Recognition

What Is Hazard Analysis?

- Methodology for the discovery of hazards
  - Rigorous
  - Systematic
  - Proactive – design out hazards early
  - Thorough – System to component coverage

- Mechanism for risk control
  - Assess risk presented by hazards
  - Determine if risk must be modified
  - Establish risk mitigation methods

> HA is the core of the System Safety approach

Why Hazards Exist

- Things fail
  - Wear, ageing, stress, manufacturing, maintenance, tolerances exceeded
- Humans error
  - Stress, lapses, drugs/alcohol, texting, training, design induced, etc.
- Designs contain flaws
  - Errors, interface problems, sneak paths, unintended functions, incorrect implementation, etc.
- Design parameters are misused or abused
- Hazard sources or assets must be used in the system
- External environment factors
- Inadequate / incorrect system integration
- Any combination of above

> When these fault conditions are not adequately countered by design safety features mishaps will result

Chapter 11 – Hazard Recognition

Types of Hazards

- Hardware Integrity
  - Critical random hardware failure or failures
    - SPF, multiple, common mode

- Systemic Integrity
  - Overall system design error or flaw
    - Poor, inadequate or insufficient design
    - The incorrect implementation of a good design
  - Inadequate / incorrect system integration
    - HW, SW, HSI, environment
  - Design fails to consider potential effect of HW or SW failures

- System
  - Fire, flood, radiation, etc.

Types of HA

- Primary HA Technique
  - A complete hazard identification technique
  - Full HA coverage
  - E.g., PHA, SHA

- Secondary HA Technique
  - An incomplete hazard identification technique
  - Not intended for full HA coverage
  - Typically have their own objective
  - Supports a primary technique
  - E.g., FMEA, FTA

Chapter 11 – Hazard Recognition

Generic HA Steps

	Steps	Tasks	Output
1	Plan HA	• Evaluate HA requirements • Select HA technique • Establish ground rules • Establish risk criteria	- HA plan - HA ground rules - Risk criteria
2	Understand System Design	• Obtain design data • Understand design	- Design questions - System boundaries
3	Acquire HA Tools	• Identify tools • Acquire tools	- Hierarchy table - TLMs & SMM
4	Identify Hazards	• Apply HA technique • Follow ground rules • Recognize hazards • Test credibility	- Hazards - Hazard causal factors - Hazard reports
5	Validate Hazards	• Peer reviews • SSWGs	- Credible hazards - Concurrence
6	Assess Risk	• Obtain failure rate data • Obtain severity data • Calculate risk	- Hazard risk
7	Mitigate Risk	• Develop hazard defenses • Establish safety design req'ts	- Safety requirements
8	Verify Mitigation	• Test result reports • Requirements verification	- Hazard closure if pass - Hazard update if failed
9	Accept Risk	• Package hazards • Obtain approvals	- Risk acceptance letter - Signatures
10	Track Hazards	• Track/Record hazards • Close hazards	- Hazard database; lessons learned - Safety case

Hazard Identification

- What To Look For
    - Energy sources (HW, chemicals, environment, etc.)
    - Hazardous or Critical :
        - Functions (fail, erroneous, timing)
        - Modes
        - Tasks
    - Critical HW Failure modes
    - Critical control functions
    - Common mode / common cause failures (HW & SW)
    - Potential human error modes that are safety-critical
    - TLM causal factors

Chapter 11 – Hazard Recognition

Key Hazard Recognition Factors

- Hazard sources (hazard assets)
- Hazard checklists
- Lessons learned
- Safety criteria
- Key failure state questions
- Good design practices
- TLMs
- SCFs
- The SMM and its overall layout
- Hazard Triangle components

Hazard Recognition Methods

- Energy sources
- Critical system functions
- Redundant items
- Expected environments
- System hierarchy table
- Hazardous assets
- Existing design safety features
- TLMs / SMM
- Hazardous subsystems
- Critical software functions
- System dependencies
- Hazard checklists
- Lessons Learned
- Design Requirements
- Hazard Organization

Chapter 11 – Hazard Recognition

HA Considerations

Key Failure State Considerations
- Fails to operate
- Fails to function
- Malfunction
- Degraded function
- Inadvertent function
- Incorrect input to function
- Incorrect output from function
- Function timing error – early, late, slow
- Unable to stop/control
- Operator confusion
- Operator misuse (reasonable)
- Out of sequence
- Erratic function

System Considerations
- Performance
- Control
- Display
- Detection
- Operator
- Computation
- Data

HA Considerations

- Materials hazardous to human health
- Chemical hazards
- Radiological hazards
- Biological hazards
- Ergonomic hazards
- Physical hazards
- Human health hazards
- Cautions, Warnings and personal protective equipment
- Information from OSHA and Environmental evaluations
- Operational requirements, constraints, and required personnel capabilities
- Human factors engineering data and reports

Chapter 11 – Hazard Recognition

System Views

System Perspective	Safety Considerations
Physical	This view involves the various architectural views that depict what the system contains and how it is constructed. It establishes subsystems, assemblies, components and the overall system equipment hierarchy. The HA should identify these physical elements and show that each was covered.
Functional	This view involves the functions that the system must perform in order to operate. All functions should be identified and evaluated by the HA. Safety-critical functions will be identified at this stage.
Operational	This view involves how the system will operate and how the user interfaces with the system, including instructions, tasks, conditions, parameters and limitations.
Software	This view the system software. Primarily, software functions are identified and evaluated by the HA. Safety-critical software functions will be identified at this stage. Functional software hazards must be also carried into the hardware and human aspects involved.

System Views (cont'd)

System Perspective	Safety Considerations
Environment	This view considers the various environments that the system will encounter (internal and external). This includes natural environments (e.g., weather, tornadoes) as well as system environments (e.g., heat, EMI).
Human	This view considers human performance in the system and the effect of potential human errors. It also includes user interfaces with the system and their potential impact on the user. Human error should be looked at from all perspectives, including how the system design can force the user to commit an error.
Organizational	This considers the organizational and management aspects affecting a hazard.

Chapter 11 – Hazard Recognition

Primary HA Steps

- Develop a list of system equipment (hardware and software)
- Develop a list of system functions
- Develop a system functional diagram
- Develop a system hierarchy table
- List overall safety concerns for system
- List design safety features existing in system design
- Identify general hazards and mishap categories
- Establish SMM
- Identify hazards using hazard recognition aids and tools
- Check hazards for wording, consistency and correctness

Typical HA Process

- Specific input, output and process is based on analysis methodology
- Focus on system HW, functions, energy sources, modes and states
- Some iteration involved between HA Types

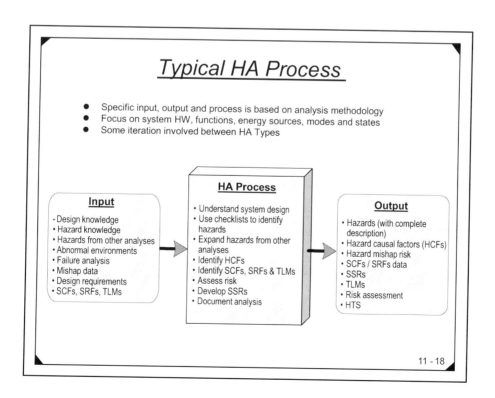

Input
- Design knowledge
- Hazard knowledge
- Hazards from other analyses
- Abnormal environments
- Failure analysis
- Mishap data
- Design requirements
- SCFs, SRFs, TLMs

HA Process
- Understand system design
- Use checklists to identify hazards
- Expand hazards from other analyses
- Identify HCFs
- Identify SCFs, SRFs & TLMs
- Assess risk
- Develop SSRs
- Document analysis

Output
- Hazards (with complete description)
- Hazard causal factors (HCFs)
- Hazard mishap risk
- SCFs / SRFs data
- SSRs
- TLMs
- Risk assessment
- HTS

Chapter 11 – Hazard Recognition

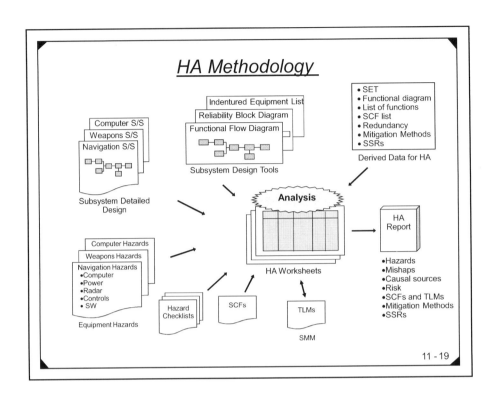

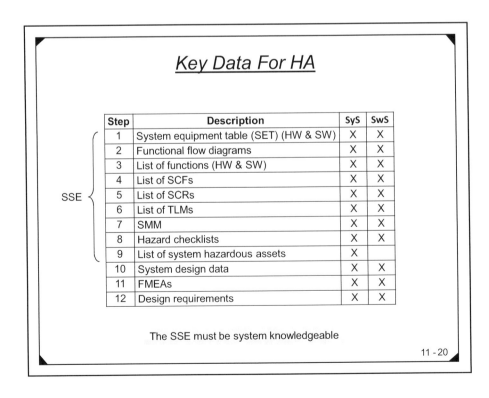

Chapter 11 – Hazard Recognition

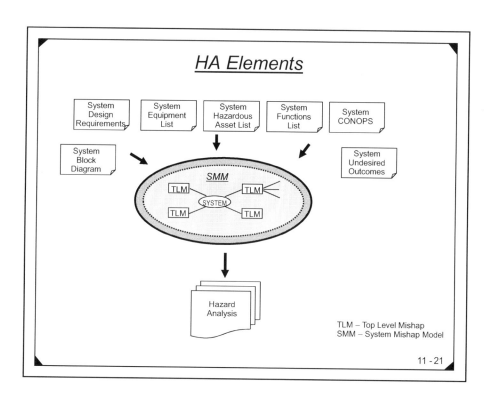

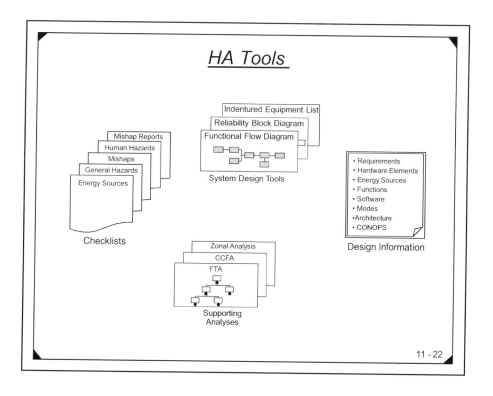

Chapter 11 – Hazard Recognition

Energy Source Checklist

1. Fuels
2. Propellants
3. Initiators
4. Explosive charges
5. Charged electrical capacitors
6. Storage batteries
7. Static electrical charges
8. Pressure containers
9. Spring-loaded devices
10. Suspension systems
11. Gas generators
12. Electrical generators
13. R. F. energy sources
14. Radioactive energy sources
15. Falling objects (gravity)
16. Catapulted objects
17. Heating devices
18. Pumps, blowers, fans
19. Rotating machinery
20. Actuating devices
21. Nuclear
22. Cryogenics

General Hazard Checklist

1. Acceleration
2. Contamination
3. Corrosion
4. Chemical dissociation
5. Electrical
    - shock
    - thermal
    - inadvertent activation
    - power source failure
6. Explosion
7. Fire
8. Heat and temperature
    - high temp
    - low temp
    - temp variations
9. Leakage
10. Moisture
    - high humidity
    - low humidity
11. Oxidation
12. Pressure
    - high
    - low
    - rapid change
13. Radiation
    - thermal
    - electromagnetic
    - ionizing
    - ultraviolet
14. Chemical replacement
15. Shock (mechanical)
16. Stress concentrations
17. Stress reversals
18. Structural damage or failure
19. Toxicity
20. Vibration and noise
21. Weather and environment
22. Gravity

Chapter 11 – Hazard Recognition

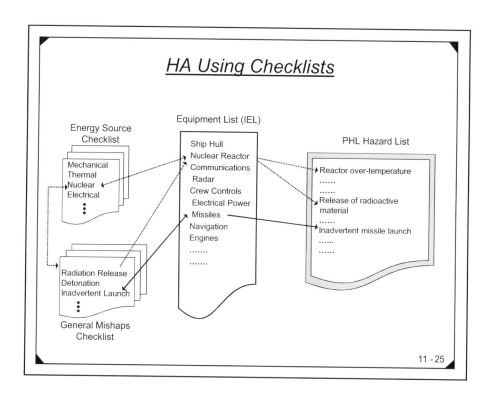

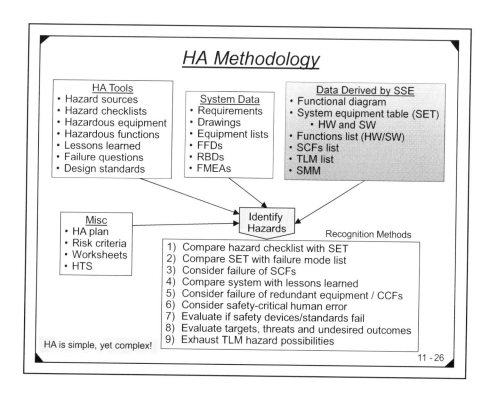

Chapter 11 – Hazard Recognition

HA Common Mistakes

- Not considering all concerns or potentially credible hazards
- Failure to document hazards identified but found to be not credible
- Not utilizing a structured approach (e.g., worksheet)
- Not collecting and utilizing common hazard source checklists
- Not researching similar systems or equipment for mishaps and lessons learned that can be applied
- Not establishing a correct list of hardware, functions and system modes/states
- Assuming the reader will understand the hazard description from a brief abbreviated statement filled with project unique terms and acronyms
- Inadequately describing the identified hazard
  - Insufficient detail, too much detail
  - Incorrect hazard effect
  - Wrong equipment indenture level
  - Not identifying all three elements of a hazard (S, M, O)

HA Common Mistakes (con't)

- Inadequately describing the HCFs
  - The identified HCF does not support the hazard
  - The HCF is not detailed enough
  - Not all of the HCFs are identified
- Inadequately/incorrectly describing the Hazard Mishap Risk
  - The risk is not stated or is incomplete
  - The hazard severity level does not support actual hazardous effects
  - The final MRI is a higher risk than the initial MRI
  - The final severity level is less than the initial severity level (sometimes possible, but not usually)
  - The hazard probability is not supported by the causal factors.
- Providing recommended hazard mitigation methods that do not address the actual HCFs
- Incorrectly closing the hazard
  - Not following a risk acceptance & closure process
  - Assuming a mitigated hazard is a closed hazard
- Failure to establish and utilize SCFs and TLMs
- Failure to establish a SMM

Chapter 11 – Hazard Recognition

Hazard Documentation

- When identifying hazards, <u>do not discount any hazard ideas</u>
- If the hazard appears not possible due to design, etc, keep it in the analysis anyway
- Show why it is not a hazard (possible design credit)
- Document for later observers who will ask, "did you consider such and such…?"
- If you do not document, you cannot show that it was indeed considered (especially 2 or 3 years later)

Good HA documentation is important (record, history, liability)

HA Rules Of Thumb

- Document everything (do not discard a hazard consideration)
- PHA will catch 90% of the system hazards
- A hazard is not eliminated unless one leg of the Hazard Triangle is eliminated
- Hazard mitigation is not the same as elimination
- No stone goes unturned (i.e., look into everything)
- The objective of HA is a safe design; this requires establishing:
  - Hazard causal factors (HS & IMs)
  - Hazard risk (likelihood & severity)

Chapter 11 – Hazard Recognition

Major HA Question

- How Do You Know When HA Is Complete?
    - When you provide <u>evidence</u> that the following have been covered:
        - All functions
        - All hardware items
        - All software modules
        - All hazardous assets
        - All environments
        - All human tasks
    - When you run out of money!! [not]

In reality you never know, but these methods and system coverage provide high confidence. Leave anything out and the confidence level goes down.

11 - 31

Summary

- HA is not rocket science, it's much more complicated
    - Identify and assess potential future events
    - Many complex factors involved
    - HA seems like a simple process, however it can easily become complex
- Hazards are primarily identified (discovered) via the HA process
- Using formal HA techniques and methods provides the best results
- HA planning and organization is key
- Trained personnel are essential (experience is beneficial)

- It is important to note that hazards can exist and occur without the presence of a hardware failure mode or software error
    - Subtle design errors
    - Requirement errors / conflicts
    - Sneak circuit paths

11 - 32

Chapter 11 – Hazard Recognition

System Safety and Reliability Analysis
Course Notes

Chapter 12
Hazard Mapping

Clifton A. Ericson II
Design Safety Solutions LLC
cliftonericson@verizon.net
540-786-3777

© C. A. Ericson II 2014

HA Problem

- False premise --- "HA is simple, anyone can do it"
- Correctly initiating a HA is critical
  - Lost time and effort spent going down the wrong rabbit hole
  - Incorrect hazards result – incorrect risk
- Driving factors
  - System complexity
  - System size
  - Lack of design information
  - Lack of training
- HA appears simple, but is actually very complex

System are large, complex and diverse – where to start can become daunting

---

Chapter 12 – Hazard Mapping

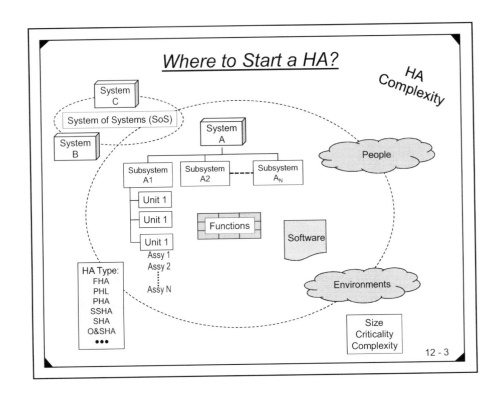

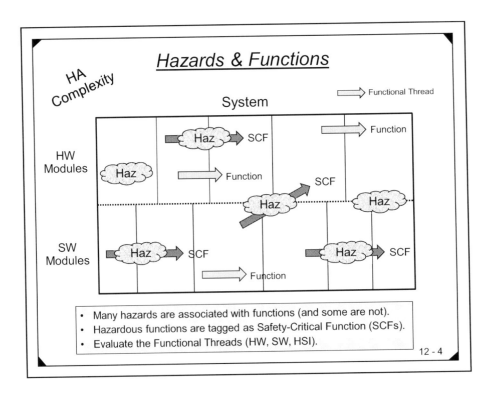

Chapter 12 – Hazard Mapping

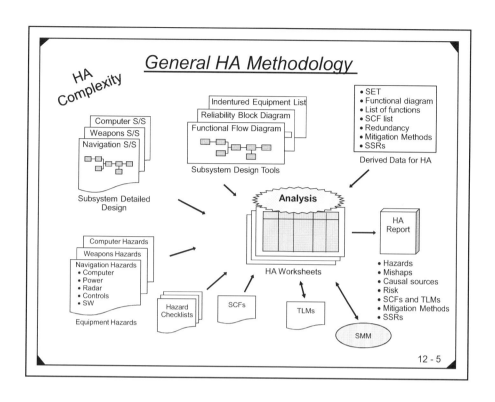

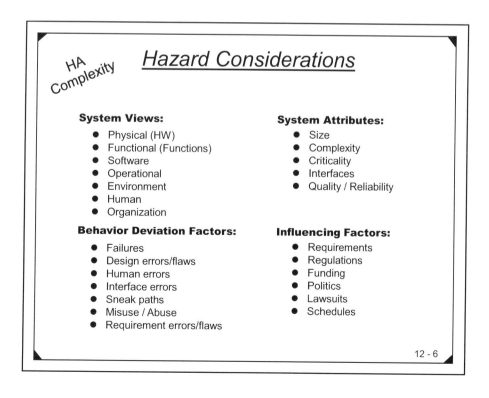

Chapter 12 – Hazard Mapping

HA Considerations

Anomaly List Considerations
- Fails to function
- Malfunction
- Degraded function
- Inadvertent function
- Incorrect input to function
- Incorrect output from function
- Function timing error – early, late, slow
- Unable to stop/control
- Operator confusion
- Out of sequence
- Erratic function

System Considerations
- Performance
- Control
- Display
- Detection
- Operator
- Computation
- Data

HA Problem - Restated

- HA is complex
- HA requires many system considerations
- HA requires intimate design knowledge
- The analyst becomes lost between the trees and the forest
- Good HA is <u>not</u> simple

- Lack of a roadmap often results in confusion and poor analyses
    - Where and how to start a HA is often critical

Chapter 12 – Hazard Mapping

HA Complexity Solution

- Simplify the process with a *roadmap*
- Utilize a system model for hazard visualization
- Employ a SMM to drive and guide the HA
- Utilize Mindmapping to develop the SMM

- A SMM is an effective tool for organizing and guiding the HA

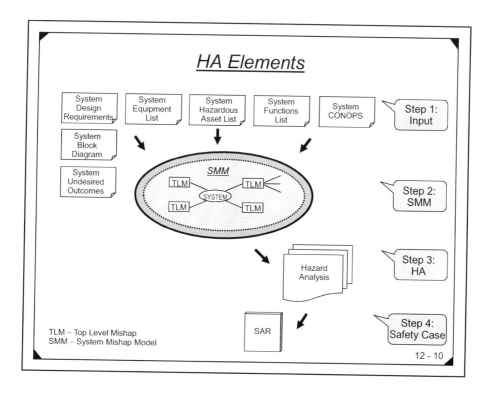

Chapter 12 – Hazard Mapping

SMM

- A SMM is the structured organization of hazards and potential mishap outcomes
- A diagram type roadmap for structured HA
- Identifies areas of potential system mishap vulnerability
- A map that links mishaps and hazards and the logical path of causal factor in between
- Organizes the major potential mishaps that a system is susceptible to

- The SMM could take different forms, such as a list, spreadsheet or logic diagram
- My recommended approach is to use a Mind Map
- Can be started with minimal design information

Top Level Mishap (TLM)

- A TLM is a generic mishap category for collecting together various hazards that share the same general outcome or type of mishap.

- A TLM is a common mishap outcome that can be caused by one or more hazards; its purpose is to serve as a collection point for of all the potential hazards that can result in the same outcome, but have different causal factors.

- TLMs provide a design safety focal point for a particular safety concern (i.e., the TLM outcome). Each contributing hazard has different initiating mechanisms or causal factors, but a common TLM outcome event. This common outcome is extracted from the hazards and used as a common TLM to unite the hazards.

- TLM "lumping" provides better visibility.

Chapter 12 – Hazard Mapping

Example TLMs

Missile System	Aircraft System	Spacecraft System
• Inadvertent missile launch • Inadvertent warhead initiation • Incorrect missile target • Self-destruct fails • Electrical injuries • Mechanical injuries • RF radiation injuries • Weapon-ship fratricide	• Controlled flight into terrain • Loss of all engines • Loss of all flight controls • Loss of landing gear • Inadv thrust reverser operation • Electrical injuries • Mechanical injuries	• Loss of oxygen • System Fire • Re-entry failure • Temperature control fails • Communications fail • Electrical injuries • Mechanical injuries

TLM

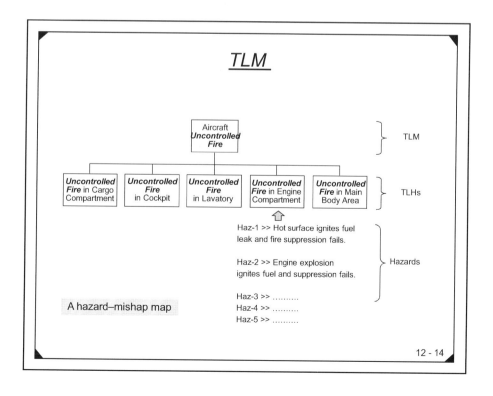

A hazard–mishap map

Haz-1 >> Hot surface ignites fuel leak and fire suppression fails.

Haz-2 >> Engine explosion ignites fuel and suppression fails.

Haz-3 >> ....
Haz-4 >> ....
Haz-5 >> ....

Chapter 12 – Hazard Mapping

System Mishap Model (SMM)

A SMM is a map of hazard–mishap relationships

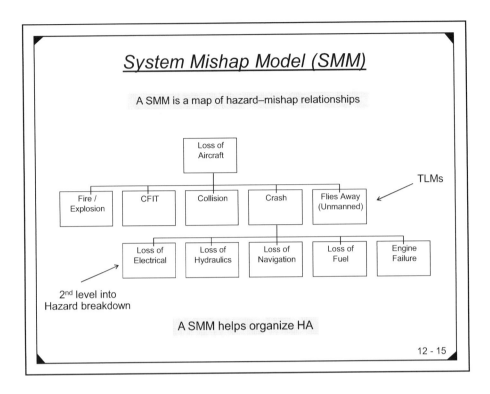

TLMs ← (Fire/Explosion, CFIT, Collision, Crash, Flies Away (Unmanned))

2nd level into Hazard breakdown

A SMM helps organize HA

12 - 15

Mindmapping

- Mindmapping (MM) is a well known approach to problem solving.
- MM is a methodology for visualizing and modeling an idea multi-dimensionally by linking the various associated thoughts and ideas that are relevant.
- It is essentially a brainstorming method for creating and organizing thought patterns.
- Developed by psychologists because it integrates the left and right spheres of the brain, which ultimately produces a better analysis result.
- Boeing created a 25-foot long mind map summarizing an aircraft engineering manual.

Free software is available

12 - 16

Chapter 12 – Hazard Mapping

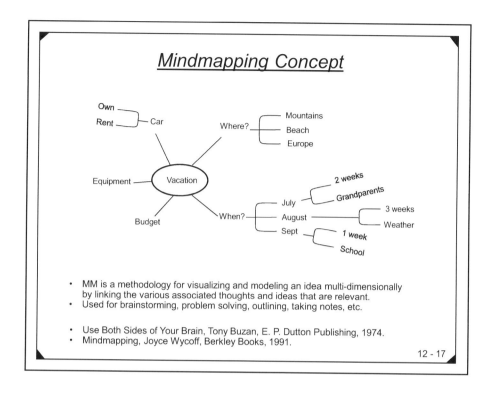

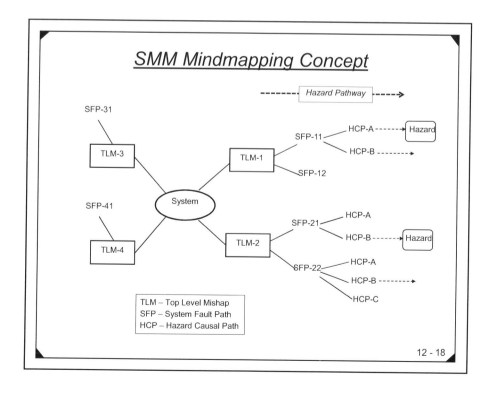

Chapter 12 – Hazard Mapping

Basic MM Steps

- Establish the central focus is a few short key words
- Place the central focus of analysis in a bubble in the center of the page
- Develop sub-topics emanating from the central focus
- Build the diagram by developing and connecting sub-topics, sub-sub-topics, etc.
- Ideas are allowed to flow freely without judgment outward from the bubble
- Key words are used to represent ideas and are expanded as desired
- Key ideas are connected with lines and color can be utilized for emphasis

SMM Process Using MM

- Identify the undesired outcomes for the system (i.e., TLMs)
- Place the TLMs on the SMM diagram (i.e., link to main bubble in mind map)
- For each TLM begin identifying the next item in the system design that would be of safety concern for the TLM
- Follow each path backwards going through functions, hardware and components until discrete hazards can be clearly identified

Chapter 12 – Hazard Mapping

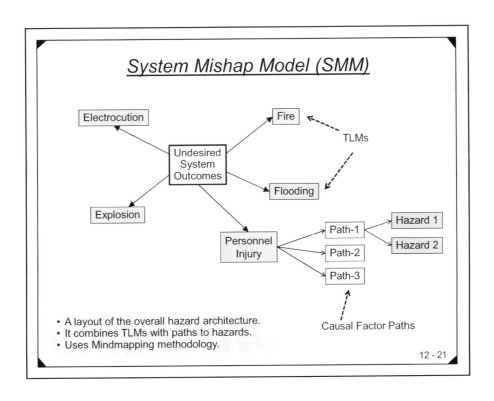

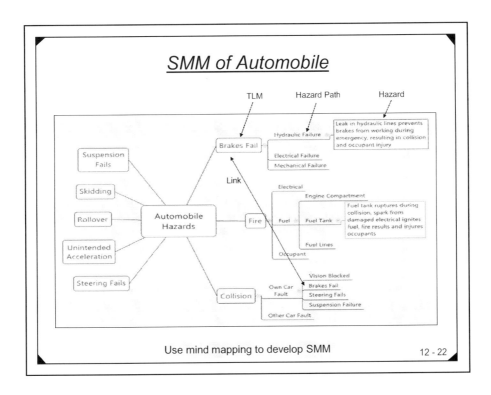

Chapter 12 – Hazard Mapping

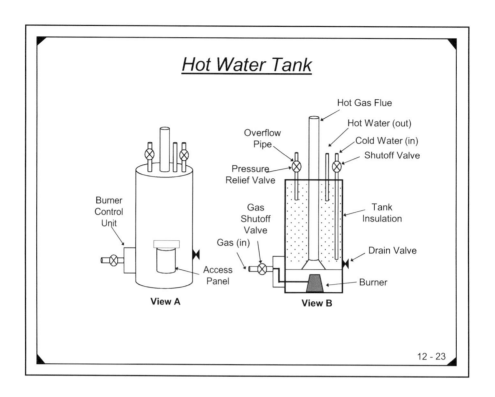

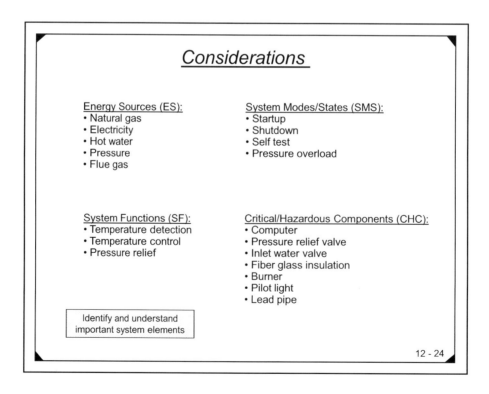

Chapter 12 – Hazard Mapping

PHL

ES: Natural gas
Gas leak occurs, gas is ignited and burns house down causing injury, death, damage.

ES: Natural gas
Gas leak occurs, occupants suffocated by vapors, causing death.

CHC: Pressure relief valve
Pressure relief valve fails, causing tank explosion, resulting in injury, death, damage.

CHC: Fiber glass insulation
Fiber glass material are exposed and cause human health hazard (HazMat).

SF: Temperature control
Temperature control fails, causing excessive high temperature water that scolds user resulting in serious burn injury.

Evaluate safety impact of system elements

SMM for Hot Water System

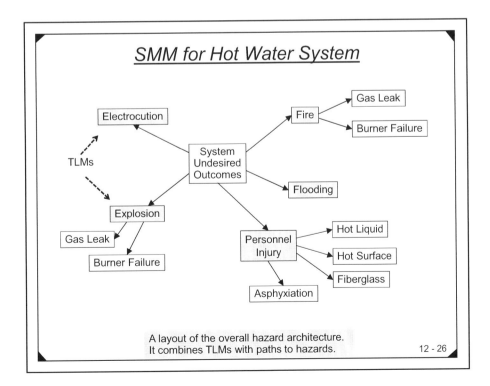

A layout of the overall hazard architecture.
It combines TLMs with paths to hazards.

Chapter 12 – Hazard Mapping

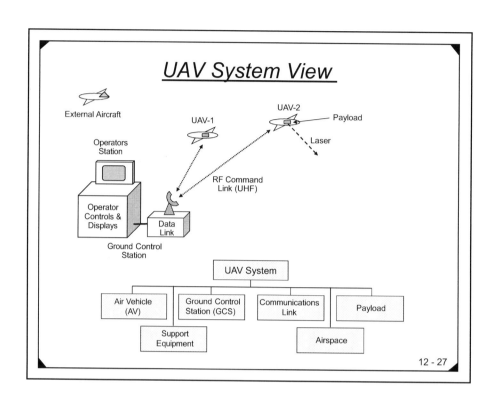

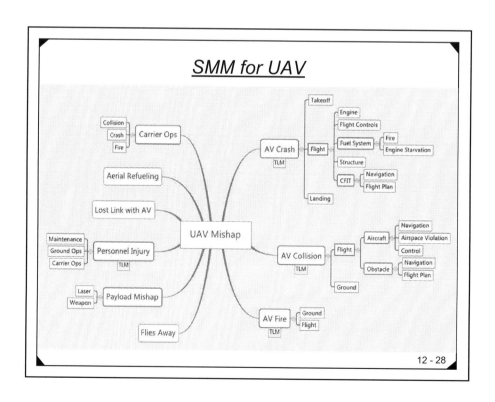

Chapter 12 – Hazard Mapping

UAV TLMs (from SMM)

1. AV Crashes
2. AV Collision
3. AV Fire/Explosion
4. Personnel Injury during Ground Operations
5. Loss of AV Ownership (flies out of range; does not return)
6. AV Returns in Unsafe State (damage, weapons armed)
7. Weapon Mishap
8. Laser Mishap

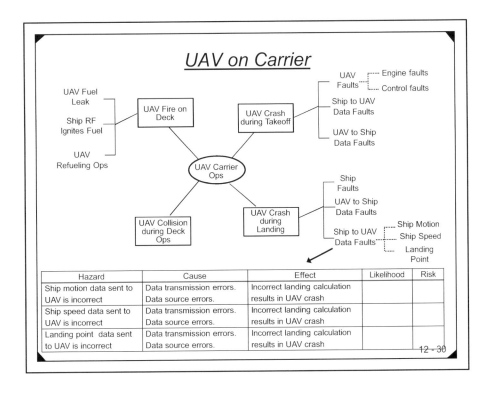

Chapter 12 – Hazard Mapping

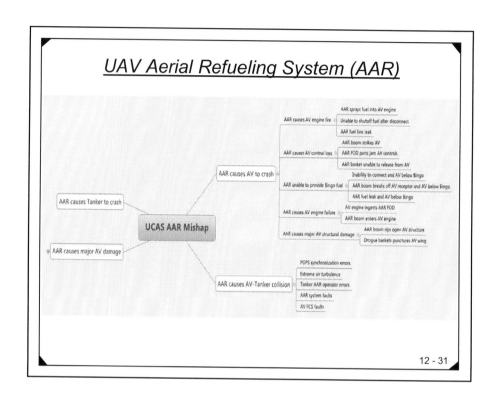

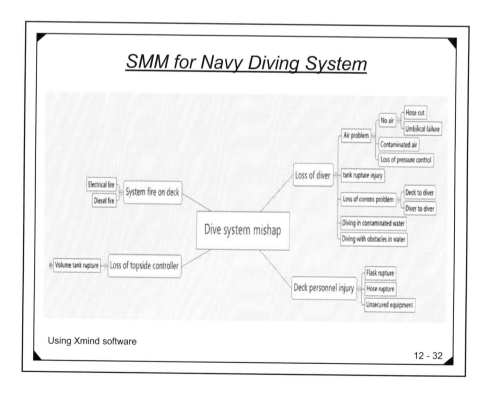

Chapter 12 – Hazard Mapping

Example #8 – Infusion Pump

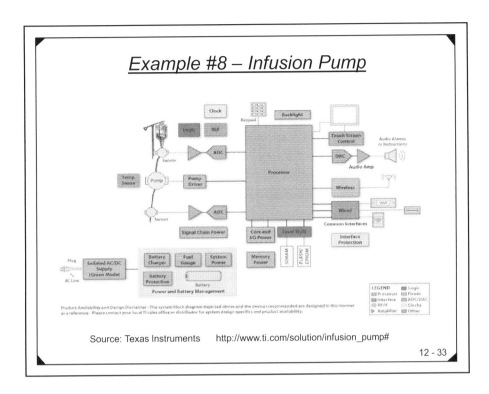

Source: Texas Instruments   http://www.ti.com/solution/infusion_pump#

Design Information

System Components:
- Processor
- Software
- Pump
- Pump Driver
- Warning Subsystem
- Fluid Sensor #1
- ADC Converter #1
- Fluid Sensor #2
- ADC Converter #2
- Touch Screen Control
- Power Input
- Backup Battery
- Wireless Interface
- Memory

System Modes/States:
- Startup
- Operation
- Shutdown
- Self test
- Warning

System Functions:
- Deliver liquid
- Shutoff liquid flow
- Control flow amount/time
- Detect abnormal conditions
- Provide warning
- Fail safe

Chapter 12 – Hazard Mapping

SMM for Infusion Pump

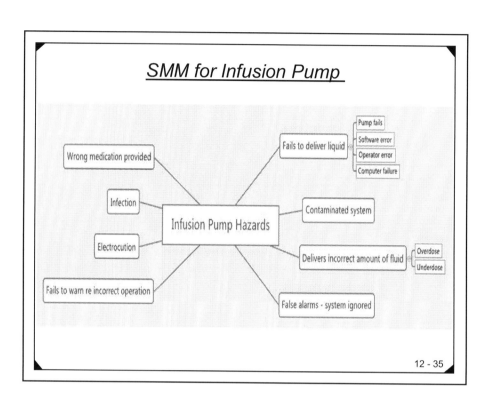

Example HA

Hazard	Effect	Cause	I-Risk	Design Safety Feature	F-Risk
Pump delivers overdose to patient due to pump failure	Patient dies	Pump part *XX* fails	1C	High reliability components.	1E
Pump delivers overdose to patient because pump is incorrectly programmed	Patient dies	• Design of setup is confusing. • Poor operator training. • Software error.	1C	System checks parameters, provides warning.	1E

Chapter 12 – Hazard Mapping

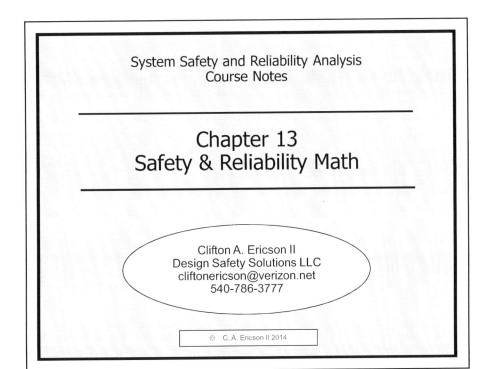

Probability

Random Phenomenon
- A phenomenon is random if individual outcomes are uncertain.
- But, there is nonetheless a regular distribution pattern in a large number of repetitions.

Probability
- Probability is a long term relative frequency.
- An order or pattern that emerges after many trials.
- For example, when tossing a coin the relative frequency is 0.5, which can be demonstrated in many trials.

Chapter 13 – Safety & Reliability Math

Probability

Sample Space
- The *set* of all possible outcomes (for a random phenomenon).
- For example, the sample space for a coin toss is S={H,T}, since there are only two possible outcomes, Heads or Tails.

Event
- A set of outcomes of random phenomenon, that is, a subset of sample space.
- For example, in a coin toss the coin landing on Heads is an event.
- The probability of an event A is denoted by $P(A)$ or P_A.
- In FTA, the failure of a component is an event, denoted by $P(A)$ or P_A.

Probability

Disjoint Events
- Events that have no outcome in common.
- Two disjoint events cannot both occur on the same trial of a random phenomenon.
- There is no overlapping of the events, they are mutually exclusive.
- The long term frequency that "one or the other" occurs is therefore the sum of their individual relative frequencies.

Complement of an Event
- If A is any event, then the event not occurring is the complement of A.
- This is denoted as A^C or $\overline{A}$.
- Note that A and $\overline{A}$ are always disjoint events, and $A + \overline{A} = 1$.
- In FTA, complement events complicates the math.

Chapter 13 – Safety & Reliability Math

Probability

Independence
- Events A and B are independent if knowing whether A occurs does not change the probability that B occurs.
- The events A and B are not disjoint.

Independent Events
- The occurrence of event A does not determine or affect the occurrence of event B.
- Knowledge about the occurrence of A give no additional information regarding the occurrence of event B.

Dependent Events
- The occurrence of event B depends upon event A occurring first.
- Event B becomes a conditional probability P(B/A).
- In FTA, dependency complicates the math.

Probability

Conditional Probability
Conditional probability is denoted as P(B/A). This is read as the probability of B occurring, given A occurs. When P(A)>0

$$P(B/A) = \frac{P(A \text{ and } B)}{P(A)}$$

Union
For two events A and B, the union is the event {A or B} that contains all the outcomes in A, in B, or in both A and B.

Intersection
For two events A and B, the intersection is the event {A and B} that contains the occurrence of both A and B.

Chapter 13 – Safety & Reliability Math

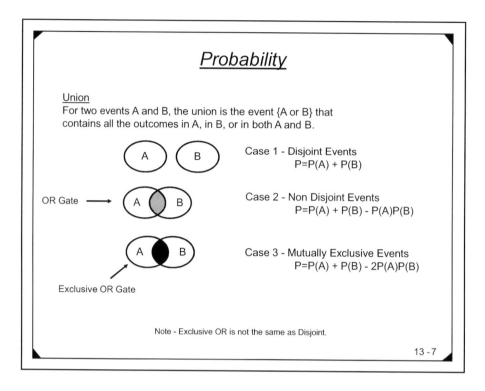

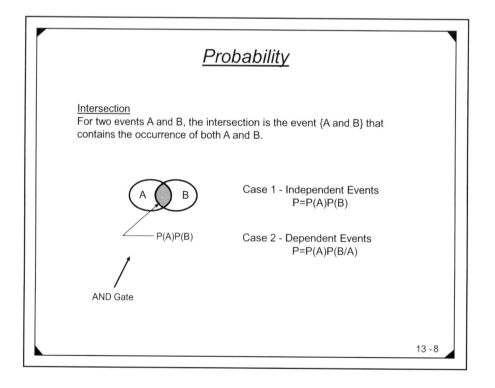

Chapter 13 – Safety & Reliability Math

Rules of Probability

[R1] The probability of an event is between 0 and 1
$$0 \le P(E) \le 1$$

[R2] If an event is certain to occur, then
$$P(E) = 100\% = 1.0$$
If an event is certain not to occur, then
$$P(E) = 0\% = 0$$

[R3] The sum of all possible outcomes for an event equals one, therefore:
$$P(E + \overline{E}) = P(E) + P(\overline{E}) = 1$$
or
$$P(\overline{E}) = 1 - P(E)$$

Rules of Probability

[R4] The General Addition Rule for the Union of Two Events:
$$P(A \text{ or } B) = P(A) + P(B) - P(A)P(B)$$

Case 1 -- Non Disjoint Events
$$P = P(A) + P(B) - P(A)P(B)$$

Case 2 -- Disjoint Events
$$P = P(A) + P(B)$$

since A and B are disjoint $P(A)P(B)=0$

Chapter 13 – Safety & Reliability Math

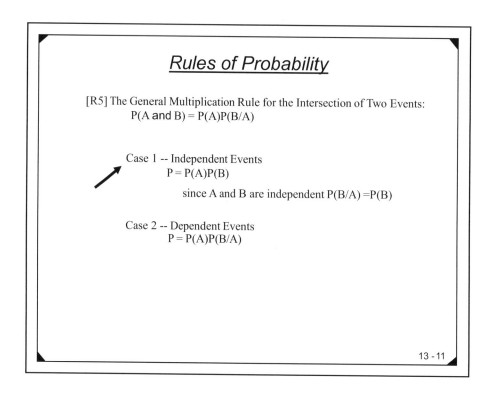

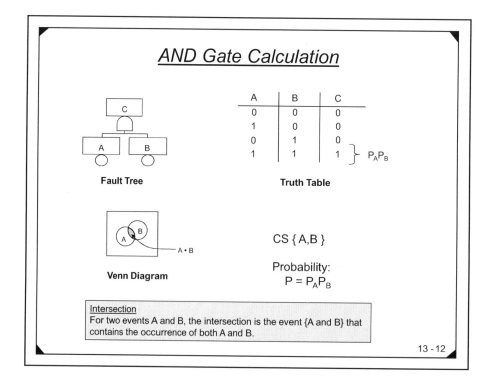

Chapter 13 – Safety & Reliability Math

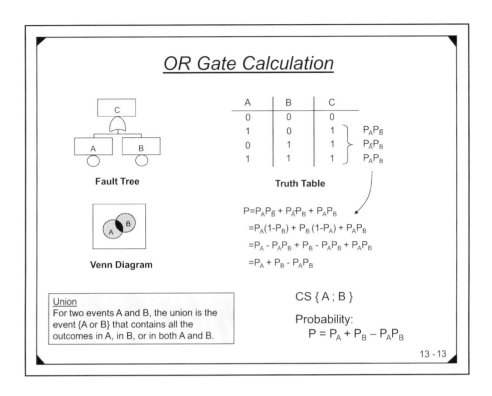

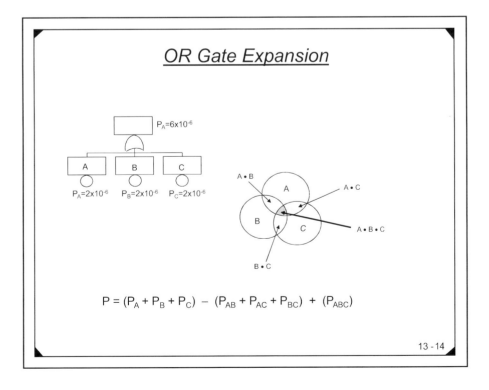

Chapter 13 – Safety & Reliability Math

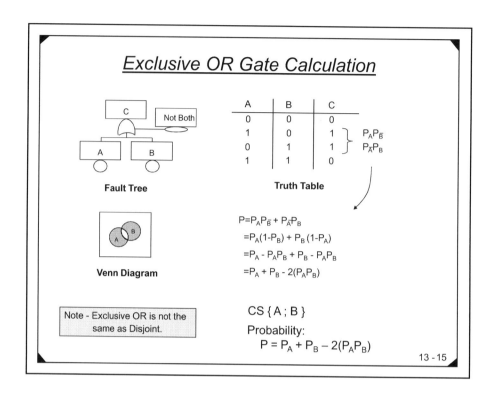
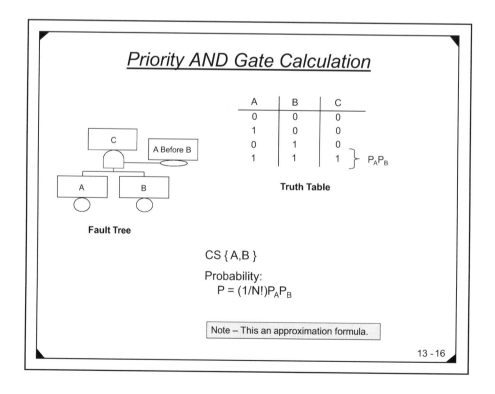

Chapter 13 – Safety & Reliability Math

Inhibit Gate

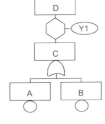

C	Y1	D	
0	P_{Y1}	0	
1	P_{Y1}	1	$P_C P_{Y1}$

Truth Table

- Both C and Y1 are necessary to cause D
- Y1 is a condition or a probability
- Pass through if condition is satisfied
- Essentially an AND gate

13 - 17

Boolean Algebra – Axioms

[A1] $ab = ba$
[A2] $a + b = b + a$ } Commutative Law

[A3] $(a + b) + c = a + (b + c) = a + b + c$
[A4] $(ab)c = a(bc) = abc$ } Associative Law

[A5] $a(b+c) = ab + ac$ } Distributive Law

13 - 18

Chapter 13 – Safety & Reliability Math

Boolean Algebra – Theorems

[T1] $a + 0 = a$
[T2] $a + 1 = 1$
[T3] $a \cdot 0 = 0$
[T4] $a \cdot 1 = a$
[T5] $a \cdot a = a$ ✓ ⎫
[T6] $a + a = a$ ✓ ⎬ Idempotent Law
[T7] $a \cdot \bar{a} = 0$
[T8] $a + \bar{a} = 1$
[T9] $a + ab = a$ ✓ ⎫
[T10] $a(a + b) = a$ ✓ ⎬ Law of Absorption
[T11] $a + \bar{a}b = a + b$

where $\bar{a}$ = not a

Example

[T5] $a \cdot a = a$ ✓ ⎫
[T6] $a + a = a$ ✓ ⎬ Idempotent Law

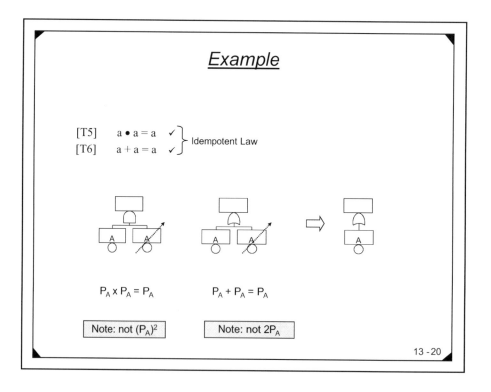

$P_A \times P_A = P_A$ $P_A + P_A = P_A$

Note: not $(P_A)^2$ Note: not $2P_A$

Chapter 13 – Safety & Reliability Math

Example

[T9] $a + ab = a$ ✓ ⎫
[T10] $a(a + b) = a$ ✓ ⎬ Law of Absorption

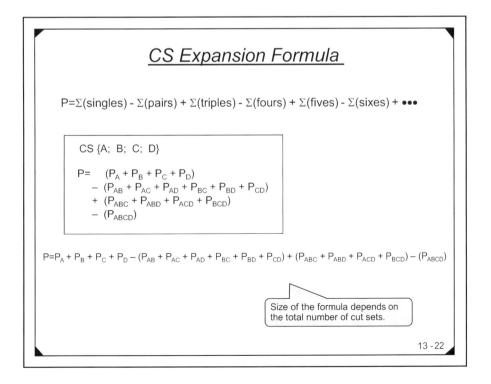

CS Expansion Formula

$P = \Sigma(\text{singles}) - \Sigma(\text{pairs}) + \Sigma(\text{triples}) - \Sigma(\text{fours}) + \Sigma(\text{fives}) - \Sigma(\text{sixes}) + \bullet\bullet\bullet$

CS {A; B; C; D}

$P = (P_A + P_B + P_C + P_D)$
$- (P_{AB} + P_{AC} + P_{AD} + P_{BC} + P_{BD} + P_{CD})$
$+ (P_{ABC} + P_{ABD} + P_{ACD} + P_{BCD})$
$- (P_{ABCD})$

$P = P_A + P_B + P_C + P_D - (P_{AB} + P_{AC} + P_{AD} + P_{BC} + P_{BD} + P_{CD}) + (P_{ABC} + P_{ABD} + P_{ACD} + P_{BCD}) - (P_{ABCD})$

Size of the formula depends on the total number of cut sets.

Chapter 13 – Safety & Reliability Math

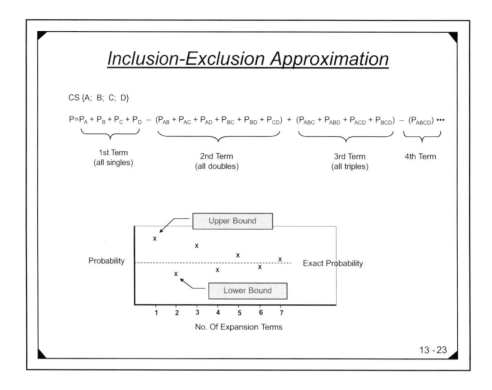

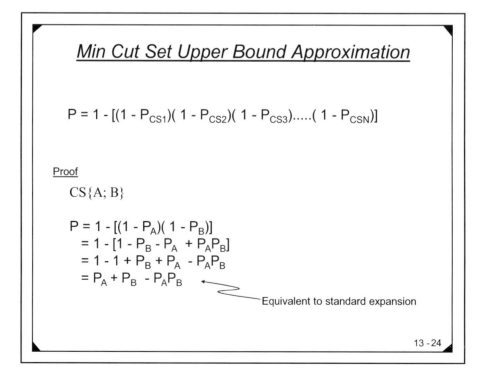

Chapter 13 – Safety & Reliability Math

Statistics

Mean = the average value of a dataset

$$\bar{x} = \text{(sum of data)} / \text{(number of data points)}$$

$$\bar{x} = \sum_{i=1}^{n} x_i / n$$

Median = midpoint value of a dataset (must arrange in order)
(if the number of data points is even, then average the two middle points)

Variance = a measure of the spread of a dataset

$$s^2 = \frac{1}{n-1} \sum_{i=1}^{n} (xi - \bar{x})^2$$

Standard deviation = the average distance of the data from the mean
= a measure of the spread of a dataset; the average variation of a value from the mean

$$s = \sqrt{s^2} = \sqrt{\frac{1}{n-1} \sum_{i=1}^{n} (xi - \bar{x})^2}$$

Example

Five people are asked how many hours of TV they watch per week

Observation	1	2	3	4	5
Value	5	7	3	38	7

Mean

$$\bar{x} = \sum_{i=1}^{n} x_i / n = (5+7+3+38+7) / 5 = 60/5 = 12 \text{ hours}$$

Median

3..5..7..7..38
        ↑
        — Median

Variance

$$s^2 = [(3-12)^2 + (5-12)^2 + (7-12)^2 + (7-12)^2 + (38-12)^2)] / (5-1) = 214$$

Standard Deviation

$$s = \sqrt{s^2} = \sqrt{214} = 14.63$$

Chapter 13 – Safety & Reliability Math

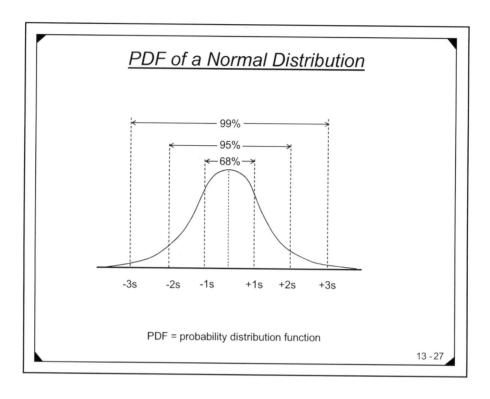

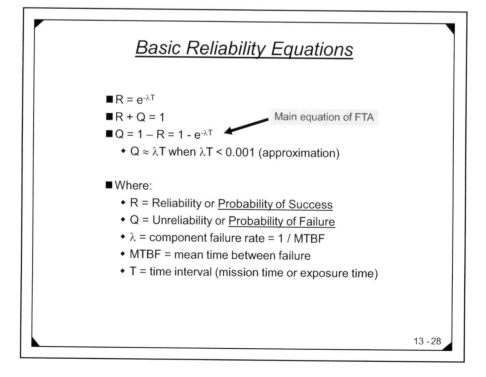

Chapter 13 – Safety & Reliability Math

System Safety and Reliability Analysis
Course Notes

Chapter 14
Failure Mode and Effects Analysis

Clifton A. Ericson II
Design Safety Solutions LLC
cliftonericson@verizon.net
540-786-3777

© C. A. Ericson II 2014

History of FMEA

- First used in the 1960's in the Aerospace industry during the Apollo missions
- In 1974, the Navy developed *MIL-STD-1629* regarding the use of FMEA
- In the late 1970's, the automotive industry was driven by liability costs to use FMEA
- Later, the automotive industry saw the advantages of using this tool to reduce risks related to poor quality

Chapter 14 – FMEA

What is an FMEA?

- An analysis methodology that examines each item in a "black box", considers how that item can fail and then determines the immediate effect

- Main products
  - Failure modes
  - Failure rates
  - Failure effect on next item and/or system
  - Seldom covers how failure cascades through the system

- Acronyms:
  - FMEA: Failure Modes and Effects Analysis
  - FMECA: Failure Modes and Effects and Criticality Analysis
    - Includes method of detection and RPN

Purpose

- The purpose is to identify the different failures and modes of failure that can occur at the component, subsystem, and system levels and to evaluate the consequences of these failures.

- Identify
  - Need for BIT
  - CSI list
  - Reliability problems
  - Safety problems
  - Need for design changes

- Used by System Safety
  - Identify hazardous failure modes
  - Provide failure rates for HA and FTA

NOTE: FMEA does not suffice for HA

Chapter 14 – FMEA

What Is A Failure Mode?

- The way in which the component, subassembly, product, input, or process could fail to perform its intended function

- Failing to meet a requirement

- Things that could go wrong

- Failure modes may be the result of upstream operations or may cause downstream operations to fail

FMEA Questions

- What can fail?
- How does it fail?
- How frequently will it fail?
- What are the effects of the failure?
- Can the failure be detected?
- What is the reliability/safety consequence of the failure?

Chapter 14 – FMEA

FMEA Worksheet

Failure Mode and Effects Analysis

Component	Failure Mode	Failure Rate	Causal Factors	Immediate Effect	System Effect	RPN

There is standard data, but no fixed worksheet

FMEA Worksheet

Failure Mode and Effects Analysis

Item	Failure Mode	Failure Rate	Causal Factors	Immediate Effect	System Effect	RPN	Method Of Detection	Current Controls	Recomm Action

Chapter 14 – FMEA

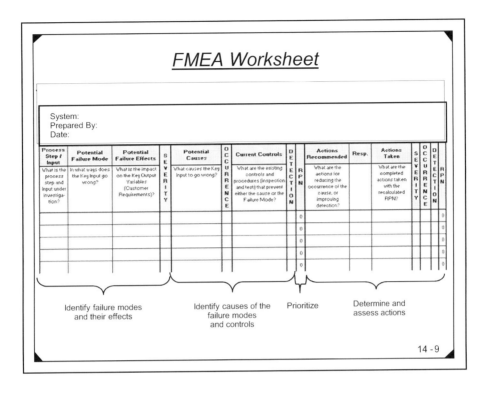

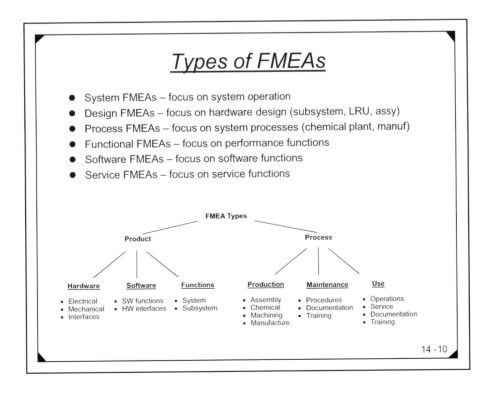

Chapter 14 – FMEA

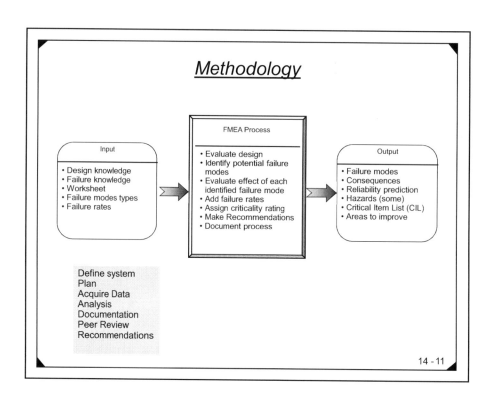

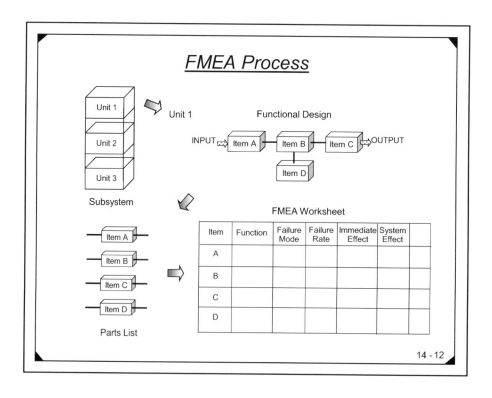

Chapter 14 – FMEA

Why Things Fail

- Fatigue/fracture
- Material removal
- Structural overload
- Radiation
- Electrical overload
- Wear (lube failure, contamination)
- Stress
- Wear (general)
- Seal failure
- Chemical attack
- Oxidation
- Ageing
- Misuse

Example Failure Modes/Categories

Modes

- Open circuit
- Short circuit
- Out of tolerance
- Leak
- Hot surface
- Bent
- Oversize/undersize
- Cracked
- Brittle
- Misaligned
- Binding
- Corroded
- Failure to operate
- Intermittent operation
- Degraded operation
- Loss of output

Categories

- Complete failure
- Partial failure (e.g., out-of tolerance)
- Intermittent failure
- Degraded function
- Unintended function
- Erroneous function

Chapter 14 – FMEA

Common Failure Modes

- Assembly
  - Missing parts
  - Damaged
  - Orientation
  - Contamination
  - Off location
- Torque
  - Loose or over torque
  - Missing fastener
  - Cross threaded
- Drilling holes
  - Missing
  - Location error
  - Over/under size
  - Angle error
- Machining
  - Too narrow
  - Too deep
  - Angle incorrect
  - Finish not to specification
  - Not cleaned
- Sealant
  - Missing
  - Wrong material applied
  - Insufficient or excessive
  - dry

Failure Modes and Effects

- The relationship between failure modes and effects is not always 1 to 1

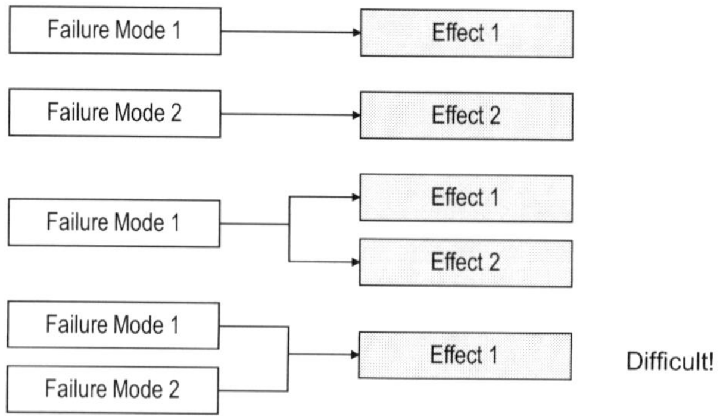

Failure Mode 1 → Effect 1

Failure Mode 2 → Effect 2

Failure Mode 1 → Effect 1
             → Effect 2

Failure Mode 1, Failure Mode 2 → Effect 1   Difficult!

Chapter 14 – FMEA

Risk Priority Number (RPN)

- RPN is the product of the severity, occurrence, and detection scores

$$\boxed{\text{Severity}} \times \boxed{\text{Occurrence}} \times \boxed{\text{Detection}} = \boxed{\text{RPN}}$$

- Severity
  1 = Not Severe, 10 = Very Severe
- Occurrence
  1 = Not Likely, 10 = Very Likely
- Detection
  1 = Likely to Detect, 10 = Not Likely to Detect

There is some question on actual value

Data Sources

- Historical Data
- Government and Commercial Data Sources
- Testing
- Research
- Analysis

Chapter 14 – FMEA

Data Sources

- Electronic Components
  - MIL-HDBK-217, Reliability Predictions of Electronic Equipment
  - Telcordia SR-332 (Bell Laboratories Bellcore)
  - PRISM (Alion System Reliability Center (SRC))
  - RIAC 217Plus (Reliability Information Analysis Center)
- B. Mechanical Components
  - NSWC Standard 98/LE1, Handbook of Reliability Prediction Procedures for Mechanical Equipment, U.S. Naval Surface Warfare Center, September 30, 1998.
  - WASH-1400 Reactor Safety Study, 1975.
- C. Human Error Rates
  - Gertman, David I. & Blackman, Harold S., Human Reliability & Safety Analysis Data Handbook, John Wiley & Sons Inc., New York, NY, 1984.
  - WASH-1400, Reactor Safety Study, 1975.
  - THERP – Technique for Human Error Rate Prediction is a methodology used for evaluating the probability of a human error occurring throughout the completion of a specific task.

Example – Gas Water Heater

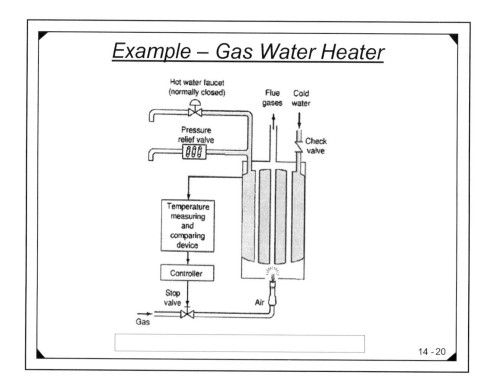

Chapter 14 – FMEA

FMEA Example for Stop Valve

	Failure Mode	Effect - Local	Effect - System
1	Fails closed	Burner off	No hot water
2	Fails open	Burner will not shut off	Overheats; release valve releases pressure; scalding
3	Fails to fully open	Burner not fully on	Water heats slowly
4	Does not respond to controller- stays open	Same as 2	Same as 2
5	Does not respond to controller- stays closed	Same as 1	Same as 1
6	Gas leaks thru valve	Burner will not shut off; burns at low level	Water may overheat
7	Gas leaks around valve	Gas leaks into room	Possible explosion, fire or gas asphyxiation

14 - 21

FMEA Example

Process or Product Name:	Hotel Service at Special Olympics														
Person Responsible:	Joe Quality					Date (Orig) _____ Revised _____									
Process Step	Key Process Input	Potential Failure Mode	Potential Failure Effect	Sev	Potential Causes	Occ	Current Controls	Det	RPN	Actions Recommended	Sev	Occ	Det	RPN
Register guest	Service Desk	Cannot Register in time	Complaints	5	Lack of language and communication skills, support of volunteers not sufficient	4	No plan on training content; training and volunteer support sufficient	3	72					
Provide Guest Services	Guest Support	Lack of barrier-free facility	Inconvenience and injury	10	Cannot provide barrier-free facility	3	Providing barrier-free facility	7	210					
Provide Meals	Food Service	Food goes bad	Disease or injury	10	Past shelf life	6	No control of raw material	8	240					
Provide Medical Service	Medical Service	Service not in time	Illness changes for worse	10	No 24 Hour service	6	12 hour service	3	180					

Source: Quality Digest/ August 2006 *Quality Service at the Special Olympics World Games*, Tang Xiaofen

14 - 22

Chapter 14 – FMEA

FMEA: A Team Tool

- A team approach is necessary.
- Team should be led by a responsible reliability engineer or similar technical individual familiar with FMEA.
- The following should be considered for team members:
  - Design Engineers
  - Process Engineers
  - Materials Suppliers
  - Customers
  - Operators
  - Reliability
  - Suppliers
  - Safety

Reasons FMEAs Fail

1. One person is assigned to complete the FMEA.
2. Members of the FMEA team are not trained in the use of FMEA, and become frustrated with the process.
3. The design or process expert is not included in the FMEA or is allowed to dominate the FMEA team.
4. FMEA team becomes bogged down with minute details of design or process, losing sight of the overall objective.
5. Rushing through identifying the failure modes to move onto the next step of the FMEA.
6. Listing the same potential effect for every failure (i.e. customer dissatisfied).
7. Not customizing the rating scales with company specific data, so they are meaningful to your company.
8. Stopping the FMEA process when the RPN's are calculated and not continuing with the recommended actions.
9. Not reevaluating the high RPN's after the corrective actions have been completed.

Chapter 14 – FMEA

Commercial Software Is Available

- Numerous types and specialized formats
- Many have free trials
  - X-FMEA Reliasoft
  - FMEA Pro-7
  - Access Data bases
  - Isograph
  - Relex

References

- MIL-STD-1629A, Procedures for Performing a Failure Mode, Effects and Criticality Analysis, 1980. http://www.fmea-fmeca.com/milstd1629.pdf
- SAE ARP5580, Recommended Failure Modes and Effects Analysis (FMEA) Practices for Non-Automobile Applications, July 2001.
- SAE ARP4761, Guidelines and Methods for Conducting the Safety Assessment Process on Civil Airborne Systems and Equipment, Appendix G – Failure Mode and Effects Analysis, 1996.
- D. H. Stamatis, Failure Mode and Effect Analysis: FMEA from Theory to Execution, 1995, Quality Press, American Society for Quality.
- R. McDermott, R. Mikulak and M. Beauregard, The Basics of FMEA, 1996, Productivity, Inc.
- SAE Standard J-1739, Potential Failure Mode and Effects Analysis in Design (Design FMEA) and Potential Failure Mode and Effects Analysis in Manufacturing and Assembly Processes (Process FMEA) and Effects Analysis for Machinery (Machinery FMEA), August 2002. http://www.fmea-fmeca.com/fmea-examples.html
- FMEA-3 Potential Failure Mode and Effects Analysis, Automotive Industry Action Group (AIAG), 3rd edition, July 2002 (equivalent of SAE J-1739).
- STUK-YTO-TR 190, Failure Mode and Effects Analysis of Software-Based Automation Systems, August 2002, Finish Radiation and Nuclear Safety Authority.
- IEC 60812, Analysis Techniques for System Reliability – Procedure for Failure Mode and Effects Analysis (FMEA), 2nd edition, 2001.

Chapter 14 – FMEA

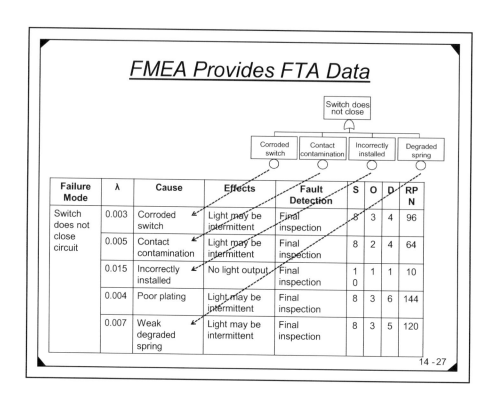

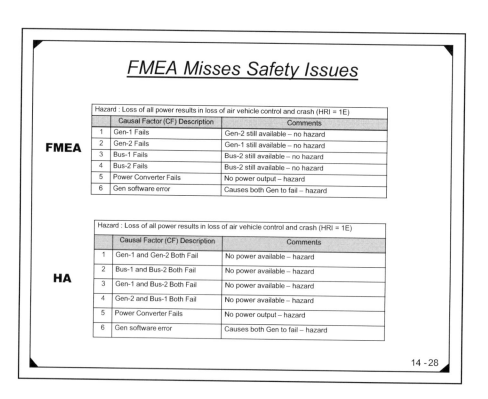

Chapter 14 – FMEA

System Safety and Reliability Analysis
Course Notes

Chapter 15
Fault Tree Analysis

Clifton A. Ericson II
Design Safety Solutions LLC
cliftonericson@verizon.net
540-786-3777

© C. A. Ericson II 2014

FTA Outline

1. FTA Introduction
2. FTA Process
3. FT Terms/Definitions
4. FT Construction
5. FT Construction Rules
6. FT of Design Models
7. FT Mathematics
8. FT Evaluation
9. FT Validation
10. FT Pitfalls
11. FT Auditing
12. FTA Application Examples
13. FTA of Example Systems
14. FT Codes

Chapter 15 – Fault Tree Analysis

--- FTA Introduction ---

Fault Tree Handbook with Aerospace Applications (updated NUREG-0492), 2002

NASA has been a leader in most technologies it has employed in its programs over the years. One of the important NASA objectives is now to add Probabilistic Risk Assessment (PRA) to its repertoire of expertise in proven methods to reduce technological and programmatic risk.

Fault Tree Analysis (FTA) is one of the most important logic and probabilistic techniques used in PRA and system reliability assessment today.

Methods to perform risk and reliability assessment in the early 1960s originated in US aerospace and missile programs. Fault tree analysis is such an example that was quite popular in the mid sixties. Early in the Apollo project the question was asked about the probability of successfully sending astronauts to the moon and returning them safely to Earth. A risk, or reliability, calculation of some sort was performed and the result was a mission success probability that was unacceptably low. This result discouraged NASA from further quantitative risk or reliability analysis until after the Challenger accident in 1986. Instead, NASA decided to rely on the use of failure modes and effects analysis (FMEA) and other qualitative methods for system safety assessments. After the Challenger accident, the importance of PRA and FTA in systems risk and reliability analysis was realized and its use at NASA has begun to grow.

Credibility

15 - 3

FTA - Description

- Tool
    - Evaluates complex systems (small to large)
    - Identify root causal factors that can result in an Undesired Event
- Model
    - Visual - displays complex cause-consequence event combinations
    - System - combines failures, errors, normal events, time, HW, SW, HE
- Analysis Technique
    - Deductive (general to the specific)
    - Provides risk assessment
        - Qualitative - cut sets
        - Quantitative - probability
- Methodology
    - Defined, structured and rigorous
    - Easy to learn, perform and follow
    - Utilizes Boolean Algebra, probability theory, reliability theory, logic
    - Proven over time

15 - 4

Chapter 15 – Fault Tree Analysis

Two Types of FTA

- Proactive FTA
  - FTA during system design development
  - Improve design by mitigating weak links in the design
  - Prevent undesired events and mishaps

- Reactive FTA
  - FTA during system operation
  - Find root causes of a mishap/accident
    - Modify the design to prevent future similar accidents

Example FT

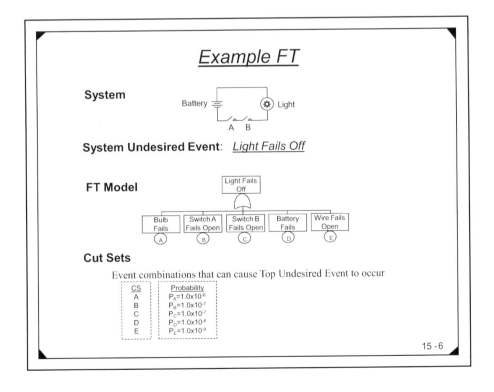

System Undesired Event: *Light Fails Off*

Cut Sets

Event combinations that can cause Top Undesired Event to occur

CS	Probability
A	$P_A = 1.0 \times 10^{-6}$
B	$P_B = 1.0 \times 10^{-7}$
C	$P_C = 1.0 \times 10^{-7}$
D	$P_D = 1.0 \times 10^{-6}$
E	$P_E = 1.0 \times 10^{-9}$

Chapter 15 – Fault Tree Analysis

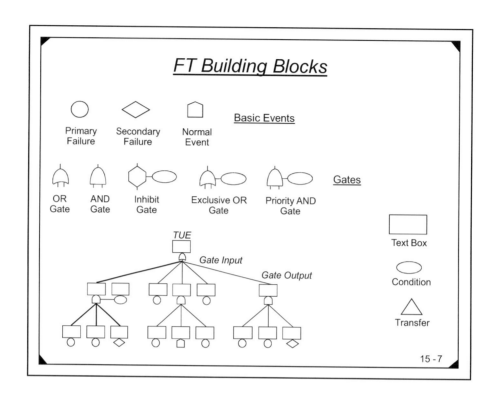

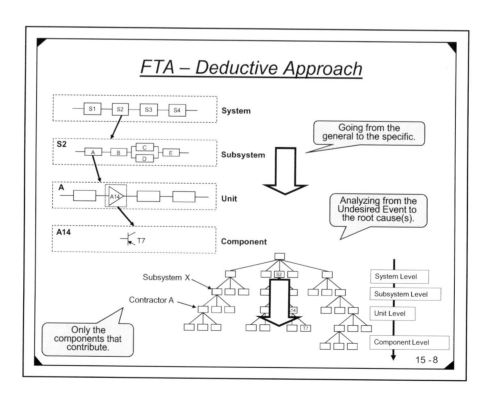

Chapter 15 – Fault Tree Analysis

FTA – Overview

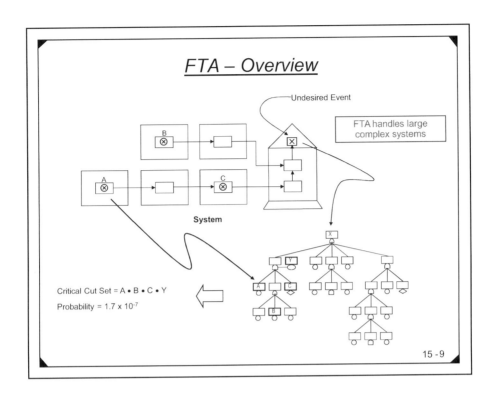

FTA Coverage

- Hardware
  - System level
  - Subsystem level
  - Component level
  - Environmental effects

- Software
  - System level control
  - Hardware/software interface

- Human Interaction
  - Human error
  - Human performance
  - Organizational structures

- Procedures
  - Operation, maintenance, assembly

- System Events
  - Failures Events
  - Normal Events
  - Environmental Effects

Chapter 15 – Fault Tree Analysis

FT Strengths

- Visual model -- cause/effect relationships
- Easy to learn, do and follow
- Models complex system relationships in an understandable manner
  - Follows paths across system boundaries
  - Combines hardware, software, environment and human interaction
  - Interface analysis - contractors, subsystems
- Probability model
- Scientifically sound
  - Boolean Algebra, Logic, Probability, Reliability
  - Physics, Chemistry and Engineering
- Commercial software is available
- FT's can provide value despite incomplete information
- Proven Technique

15 - 11

Why Do A FTA?

- Root Cause Analysis
  - Identify all relevant events and conditions leading to Undesired Event
  - Determine parallel and sequential event combinations
  - Model diverse/complex event interrelationships involved

- Risk Assessment
  - Calculate the probability of an Undesired Event (level of risk)
  - Identify safety critical components/functions/phases
  - Measure effect of design changes

- Design Safety Assessment
  - Demonstrate compliance with requirements
  - Shows where safety requirements are needed
  - Identify and evaluate potential design defects/weak links
  - Determine Common Mode failures

15 - 12

Chapter 15 – Fault Tree Analysis

When Is A FTA Necessary?

- Required by customer
- Required for certification (demonstrate compliance)
- Necessitated by high risk product
- Accident/incident/anomaly investigation
- To make a detailed safety case for safety critical system
- To evaluate corrective action or design options
- Need to evaluate criticality, importance, probability and risk
- Need to know root cause chain of events
- To evaluate the effect of safety barriers
- Determine best location for safety devices (weak links)
- Evaluate Common-Cause faults

FTA is not for every hazard' primarily for SC hazards

Example FTA Applications

- Evaluate inadvertent arming and release of a weapon
- Calculate the probability of a nuclear power plant accident
- Evaluate an industrial robot going astray
- Calculate the probability of a nuclear power plant safety device being unavailable when needed
- Evaluate inadvertent deployment of jet engine thrust reverser
- Evaluate the accidental operation and crash of a railroad car
- Evaluate spacecraft failure
- Calculate the probability of a torpedo striking target vessel
- Evaluate a chemical process and determine where to monitor the process and establish safety controls

Chapter 15 – Fault Tree Analysis

FT Weaknesses

- Knowing too little (about FT's) can be dangerous
  - Simplicity is seductive (easy to go beyond knowledge)
- Evaluates only hard failures (not intermittent failures)
- Independent events only
- Limited sequential capability
- Repair is more complex process
- No logic loops

FTA Misconceptions

- Hazard Analysis or Root Cause Analysis?
  - Meets definition of hazard analysis
  - Normally used for root cause analysis and PRA
- Not an FMEA
  - FMEA is bottom up single thread analysis
- Not an Un-Reliability Analysis
  - System Integrity vs. Availability
  - Not an inverse Success Tree
- Not a model of all system failures
  - Only includes those failures pertinent to the top Undesired Event
- Not 100% fidelity – model of reality only
  - Estimate, not an exact duplicate
  - Perception of reality

Chapter 15 – Fault Tree Analysis

FTA Criticisms

- It's too difficult for an outside reviewer to know if a FT is complete
- The correctness of a tree cannot be verified
- FTA cannot handle timing and sequencing
- FTA failure data is questionable
- Can quantitative results be trusted?
- What is the uncertainty in quantitative results?
- FTs become too large, unwieldy and time consuming
- FTs are subjective
- FTA can become the goal rather than the solution
- Different analysts sometimes produce different FTs of the same system – so one must be wrong

Two Equivalent FTs

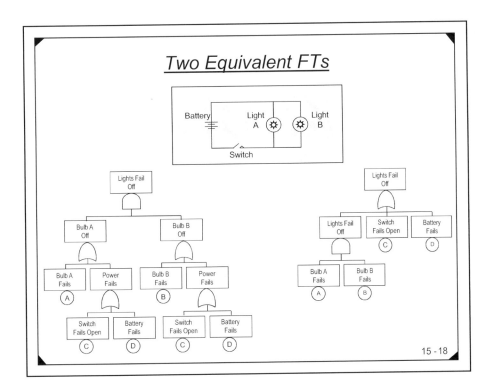

Chapter 15 – Fault Tree Analysis

FTA Misuse

- Manipulating the tree structure to obtain desired results
- Manipulating the data to obtain desired results
- Sloppy and erroneous analysis resulting in incorrect model
- Incorrect computer algorithms or approximations
- Analyzing the wrong problem
- Incorrect application

FTA Historical Stages

- H. Watson of Bell Labs, along with A. Mearns, developed the technique for the Air Force for evaluation of the Minuteman Launch Control System, circa 1961
- Recognized by Dave Haasl of Boeing as a significant system safety analysis tool (1963)
- First major use when applied by Boeing on the entire Minuteman system for safety evaluation (1964 – 1967, 1968-1999)
- The first technical papers on FTA were presented at the first System Safety Conference, held in Seattle, June 1965
- Boeing began using FTA on the design and evaluation of commercial aircraft, circa 1966
- Boeing developed a 12-phase fault tree simulation program, and a fault tree plotting program on a Calcomp roll plotter
- Adopted by the Aerospace industry and Nuclear Power Industry
- High quality FTA commercial codes developed that operates on PCs

Chapter 15 – Fault Tree Analysis

Reference Books

- Reliability and Fault Tree Analysis, Conference On Reliability And Fault Tree Analysis; UC Berkeley; SIAM Pub, R. E. Barlow & J. B. Fussell & N. D. Singpurwalla, 1975.
- → NUREG-0492, Fault Tree Handbook, N. H. Roberts, W. E. Vesely, D. F. Haasl & F. F. Goldberg, 1981.
- IEC 1025, Fault Tree Analysis, International Electrotechnical Commission, 1990.
- Reliability and Risk Assessment, Longman Scientific & Technical, 1993, J. D. Andrews & T. R. Moss, 1993.
- Probabilistic Risk Assessment and Management for Engineers and Scientists, E. J. Henley & H. Kumamoto, IEEE Press (2nd edition), 1996.
- NASA (no number), Fault Tree Handbook with Aerospace Applications, August 2002.
- NASA (no number), Probabilistic Risk Assessment Procedures Guide for NASA Managers and Practitioners, August 2002.
- Hazard Analysis Techniques for System Safety, C. A. Ericson, John Wiley & Sons, 2005, Chapter 11.
- Fault Tree Analysis Primer, C.A. Ericson, CreateSpace, 2012

15 - 21

FTA Timeline

- Design Phase
  - FTA should start early in the program
  - The goal is to influence design early, before changes are too costly
  - Update the analysis as the design progresses
  - Each FT update adds more detail to match design detail
  - Even an early, high level FT provides useful information
- Operations Phase
  - FTA during operations for root cause analysis
  - Find and solve problems (anomalies) in real time

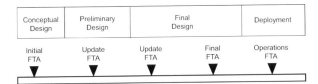

Ref: SAE ARP-4761

15 - 22

Chapter 15 – Fault Tree Analysis

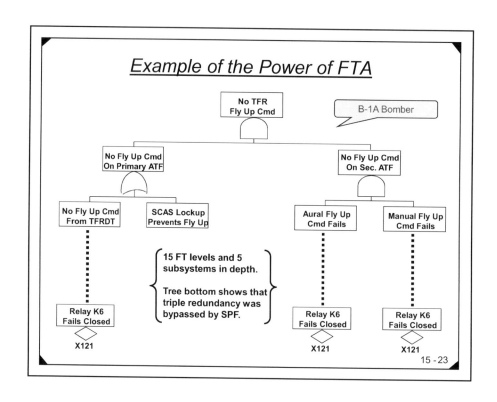

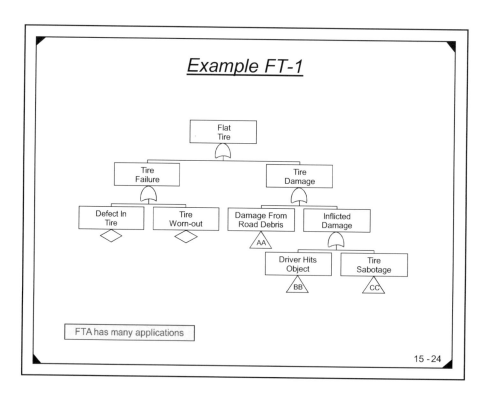

Chapter 15 – Fault Tree Analysis

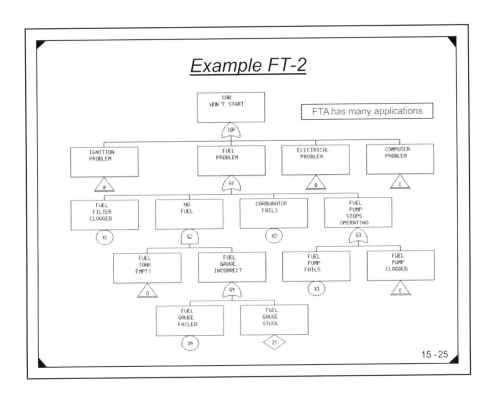

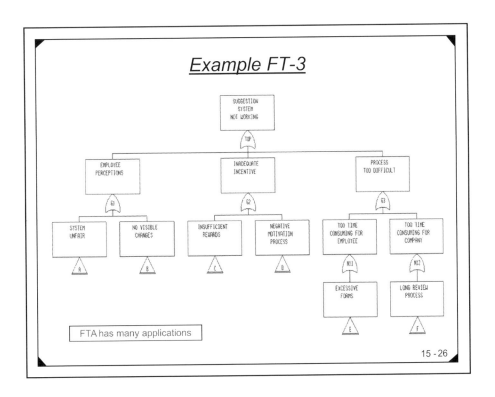

Chapter 15 – Fault Tree Analysis

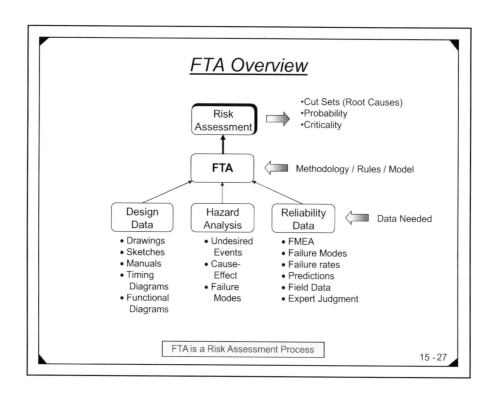

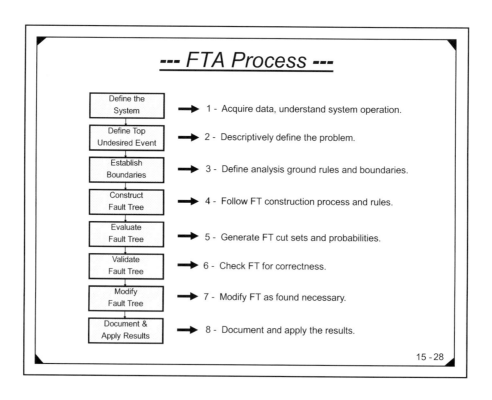

Chapter 15 – Fault Tree Analysis

Step 1 – Define The System

- Obtain system design information
  - Drawings, schematics, procedures, timelines
  - Failure data, exposure times
  - Logic diagrams, block diagrams, IELs

- Know and understand
  - System operation
  - System components and interfaces
  - Software design and operation
  - Hardware/software interaction
  - Maintenance operation
  - Test procedures

> Guideline -- If you are unable to build block diagram of the system, your understanding may be limited.

Step 2 – Define The Top Undesired Event

- Purpose
  - The analysis starts here, shapes entire analysis
  - Very important, must be done correctly
- Start with basic concern
  - Hazard, requirement, safety problem, accident/incident ⇐ Source
- Define the UE in a long narrative format
- Describe UE in short sentence
- Test the defined UE
- Determine if UE is achievable and correct
- Obtain concurrence on defined UE

Chapter 15 – Fault Tree Analysis

Example Top UE's

- Inadvertent Weapon Unlock
- Inadvertent Weapon Release
- Incorrect Weapon Status Signals
- Failure of the MPRT Vehicle Collision Avoidance System
- Loss of All Aircraft Communication Systems
- Inadvertent Deployment of Aircraft Engine Thrust Reverser
- Offshore Oil Platform Overturns During Towing
- Loss of Auto Steer-by-wire Function

Top UE Approaches

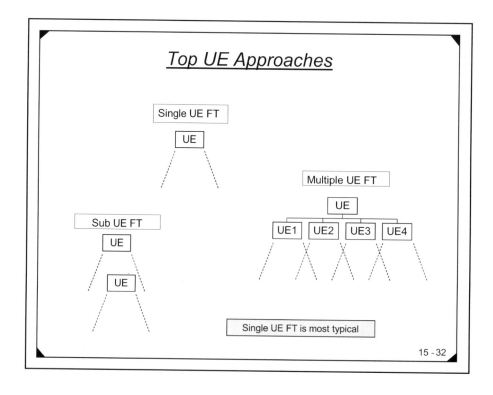

Single UE FT is most typical

Chapter 15 – Fault Tree Analysis

Undesired Event Example

System: Ship

No.	UE Candidates	Comments
1	Ship sinks	No, too broad and vague.
2	Ship runs aground	Possible, but very broad.
3	Ship communications fail	OK.
4	One engine fails	OK.
5	Both engines fail	OK.
6	Steering system fails	OK.
7	All the passengers die	No, too broad and vague.

Step 3 – Establish Boundaries

- Define the analysis ground rules
- Define assumptions
- Bound the overall problem
- Obtain concurrence
- Document the ground rules, assumptions and boundaries

Boundary Factors
- System performance – areas of impact
- Size – depth and detail of analysis
- Scope of analysis – what subsystems and components to include
- System modes of operation – startup, shutdown, steady state
- System phase(s)
- Available resources (i.e., time, dollars, people)
- Resolution limit (how deep to dig)
- Establish level of analysis detail and comprehensiveness

Chapter 15 – Fault Tree Analysis

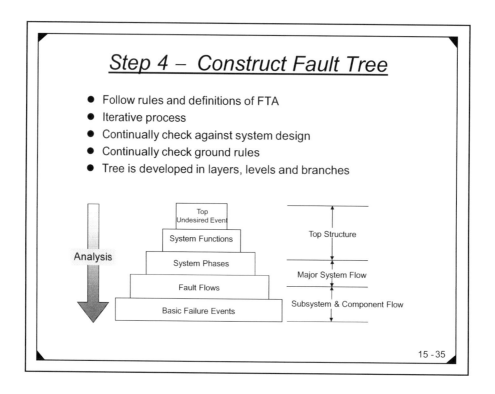

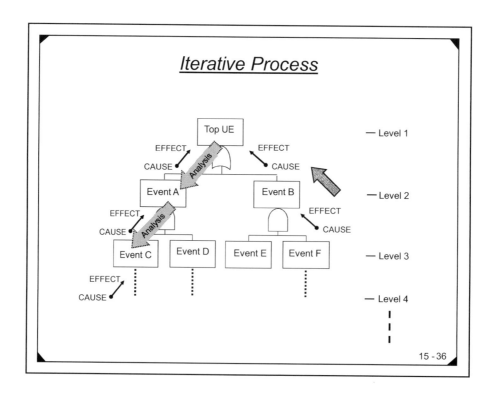

Chapter 15 – Fault Tree Analysis

Step 5 – Evaluate Fault Tree

- Qualitative Analysis
    - Generate Cut Sets
    - Verify correctness of Cut Sets
    - Evaluate Cut Sets
- Quantitative Analysis
    - Apply failure data to tree events
    - Compute tree probability
    - Compute Importance Measures
- Perform Risk Analysis
    - Criticality, Severity, Sensitivity, Probability

Generate FT results and interpret the findings

Basic Evaluation Methods

- Manual
    - possible for small/medium noncomplex trees
- Computer
    - Required for large complex trees
    - Two approaches
        - Analytical
        - Simulation
- Methods
    - Cut Set computation
        - Boolean reduction
        - Algorithms (eg, MOCUS, MICSUP)
        - Binary Decision Diagram (BDD)
    - Probability computation
        - Boolean reduction
        - Approximations

Chapter 15 – Fault Tree Analysis

Step 6 – Validate Fault Tree

- Verify the FT is correct and accurate (Objective)
    - Check FT for errors
    - Ensure correctness

- Product Quality Assurance (Purpose)
    - Management assurance
    - Customer assurance
    - Ensure quality of results

Methods For Validating The FT

- Tree reality check (are CSs correct)
- Probability reasonableness test (is tree probability reasonable based on CSs)
- Success tree inversion
- Gate check
- UE check
- Check against definitions and boundaries
- Review of failure data
- Metrics check
- Peer review
- MOE check
- Intuition check
- Logic Loop check

Chapter 15 – Fault Tree Analysis

Step 7 – Modify Fault Tree

- Modify FT when design changes are proposed/incorporated

- Make changes in FT structure as found necessary from validation
  - Validation results
  - Risk analysis results
  - Better system knowledge

- Features that can be modified
  - Tree logic
  - Tree events
  - Event failure rates

Step 8 – Document & Apply Results

- Document the study
  - Customer product (in-house or external)
  - Historical record
  - May need to update FTA some day for system upgrades
  - May need to reference the FTA study for other projects
  - Adds credibility

- Apply FTA Results
  - Interpret results
  - Present the results (using the document)
  - Make design recommendations
  - Follow-up on recommendations

Chapter 15 – Fault Tree Analysis

Document Outline Example

- Documentation should include:
  - Purpose of FTA
  - FTA scope and boundaries
  - FTA ground rules
  - System description
  - Fault trees
  - FT failure data and <u>sources</u>
  - Design data references (drawings, drawing numbers, pictures)
  - Description of FT tools used
  - FT validation
  - FT evaluation
  - FTA conclusions and recommendations

Documentation is important even for small projects

15 - 43

Summary – FTA Process

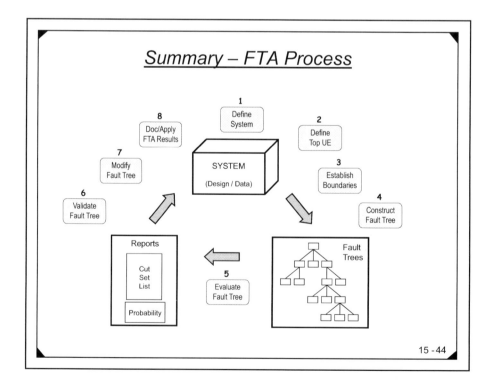

15 - 44

Chapter 15 – Fault Tree Analysis

--- FTA Terms / Definitions ---

- FT Event
  - A basic failure event on the FT
  - A normally occurring event on the FT

- FT Node
  - Any gate or event on the FT

- FT Undesired Event
  - The hazard or problem of concern for which the root cause analysis is necessary
  - The top node or event on the FT
  - The starting point for the FT analysis

Basic Fault Tree Symbols

Symbol	Action	Description	
▭	Text Box	Contains the text for a tree node	Tree Node

Symbol	Action	Description	
◯	Primary Failure	Basic primary component failure mode	Basic Events
◇	Secondary Failure	a) Secondary component failure mode b) Event that could be further expanded	
⬠	Normal Event	An event that is normally expected to occur	

Chapter 15 – Fault Tree Analysis

Basic Events (BEs)

- Failure Event
  - Primary Failure - basic component failure (circle)
  - Secondary Failure - failure caused by external force (diamond)
- Normal Event
  - An event that describes a normally expected system state
  - An operation or function that occurs as intended or designed, such as "Power Applied At Time T1"
  - The Normal event is usually either On or Off, having a probability of either 1 or 0
  - House symbol

> The BE's are where the failure rates and probabilities enter the FT

15 - 47

Event Symbol Examples

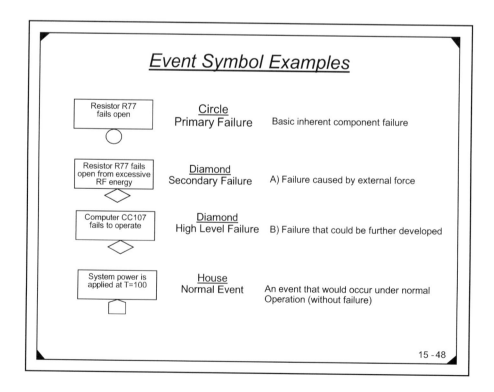

15 - 48

Chapter 15 – Fault Tree Analysis

Gate Events (GEs)

- A logic operator combining input nodes

- Five basic logic operator types
  - AND, OR, Inhibit, Priority AND and Exclusive OR
  - Additional types do exist, but usually not necessary

- Represents a fault state that can be further expanded

Gate Symbols

Symbol	Action	Description
	OR Gate	The output occurs only if at least one of the inputs occur
	AND Gate	The output occurs only if all of the inputs occur together
	Inhibit Gate	The output occurs only if the input event occurs and the attached condition is satisfied
	Exclusive OR Gate	The output occurs only if at least one of the inputs occurs, but not both
	Priority AND Gate	The output occurs only if all of the inputs occur together, but in a specified sequence (input 1 must occur before 2)

Chapter 15 – Fault Tree Analysis

Condition Events (CEs)

- A condition attached to a gate event
- It establishes a condition that is required to be satisfies in order for the gate event to occur

Symbol	Action	Description
⬭	Condition Event	A conditional restriction or an event probability

15 - 51

Transfer Event (TE)

- Indicates a specific tree branch (subtree)
- A pointer to a tree branch
- A Transfer only occurs at the Gate Event level
- Represented by a Triangle
- The Transfer is for several different purposes:
  - Starts a new page (for FT prints)
  - It indicates where a branch is used numerous places in the same tree, but is not repeatedly drawn (Internal Transfer)
  - It indicates an input module from a separate analysis (External Transfer)

△ In   △ Out

15 - 52

Chapter 15 – Fault Tree Analysis

Example Of FT Terminology

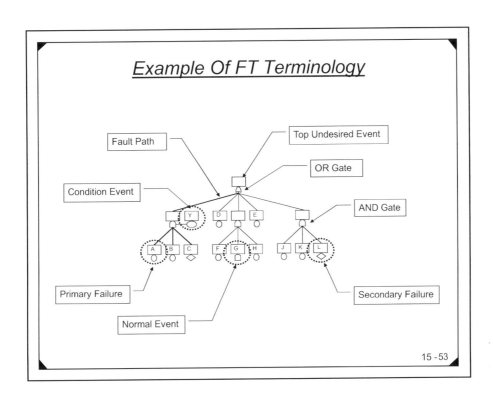

OR Gate

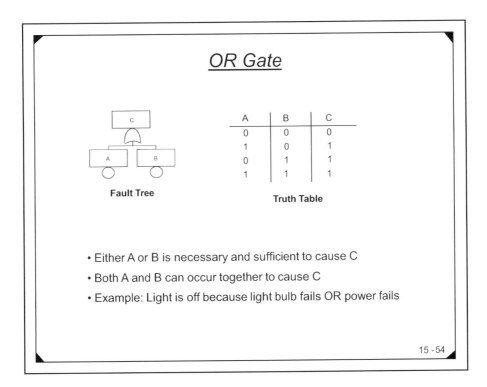

- Either A or B is necessary and sufficient to cause C
- Both A and B can occur together to cause C
- Example: Light is off because light bulb fails OR power fails

Chapter 15 – Fault Tree Analysis

OR Gate Example

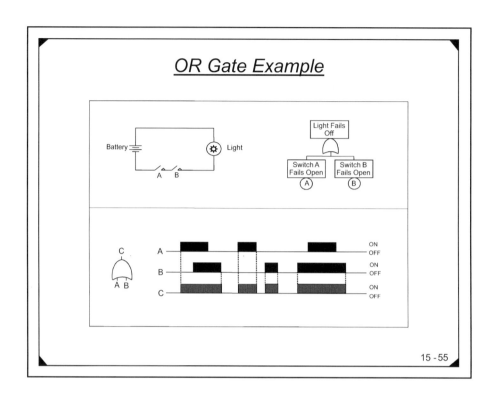

OR Gate

- Causality passes through an OR gate
  - Inputs are identical to the output, only more specifically defined (refined) as to cause
  - The input faults are never the cause of the output fault
    - Passes the cause through
    - Not a cause-effect relationship

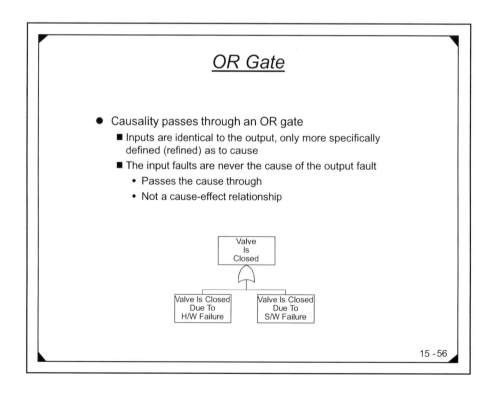

Chapter 15 – Fault Tree Analysis

AND Gate

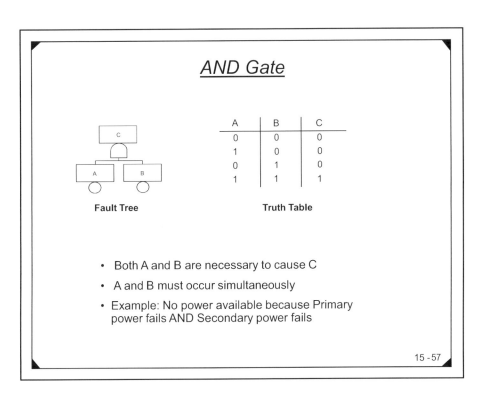

- Both A and B are necessary to cause C
- A and B must occur simultaneously
- Example: No power available because Primary power fails AND Secondary power fails

AND Gate Example

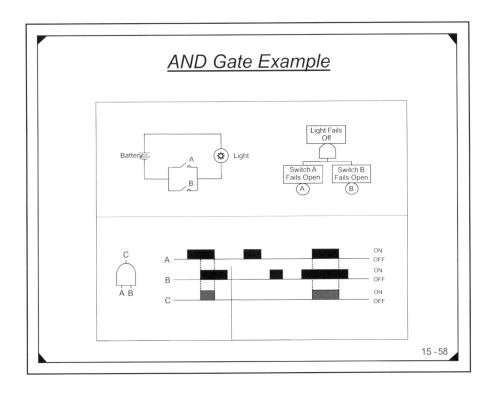

Chapter 15 – Fault Tree Analysis

AND Gate

- Specifies a <u>causal</u> relationship between the inputs and the output
  - Causality is created at the AND gate
  - The input faults collectively represent the cause of the output fault
  - Implies nothing about the antecedents of the input faults

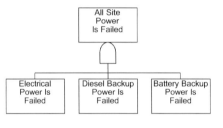

Exclusive OR Gate

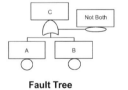

Fault Tree

A	B	C
0	0	0
1	0	1
0	1	1
1	1	0

Truth Table

- Either A or B is necessary and sufficient to cause C
- But, both A and B cannot occur together (at same time)
- Only allow two inputs (cascade down for more ExOR inputs)
- Example: Relay is energized OR Relay is de-energized, but not both

---

Chapter 15 – Fault Tree Analysis

Priority AND Gate

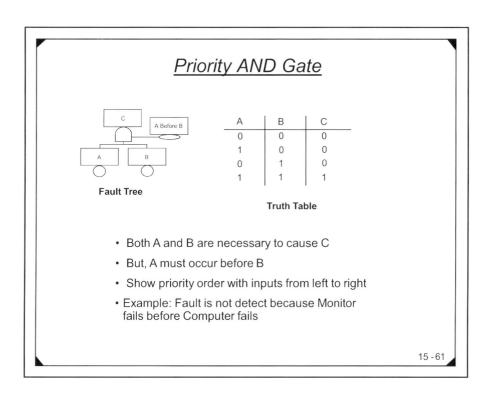

- Both A and B are necessary to cause C
- But, A must occur before B
- Show priority order with inputs from left to right
- Example: Fault is not detect because Monitor fails before Computer fails

15 - 61

Priority AND Gate

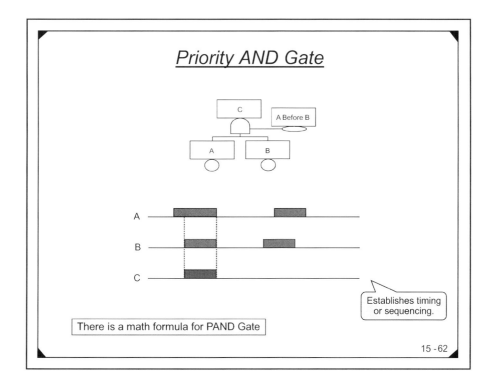

Establishes timing or sequencing.

There is a math formula for PAND Gate

15 - 62

Chapter 15 – Fault Tree Analysis

Inhibit Gate

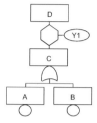

Effectively an AND gate

- Both C and Y1 are necessary to cause D
- Y1 is a condition or a probability
- Pass through if condition is satisfied
- Example: Ignition temperature is present, given faults cause overtemp AND probability that 700 degrees is reached

Transfer Symbols

Symbol	Action	Description
△	Internal Transfer	Indicates the start of a subtree branch, internal to present FT
△ (filled base)	External Transfer	Indicates the start of a subtree branch, external to present FT
▽	Similar Transfer	Indicates the start of a subtree branch that is similar to another one, but with different hardware

- The transfer is a <u>Pointer</u> to a tree branch.
- Helps to partition trees when they become large and unwieldy.

Chapter 15 – Fault Tree Analysis

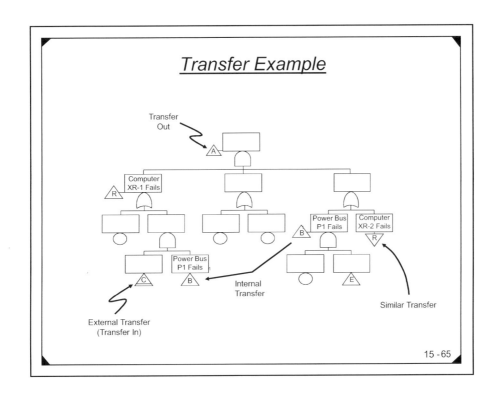

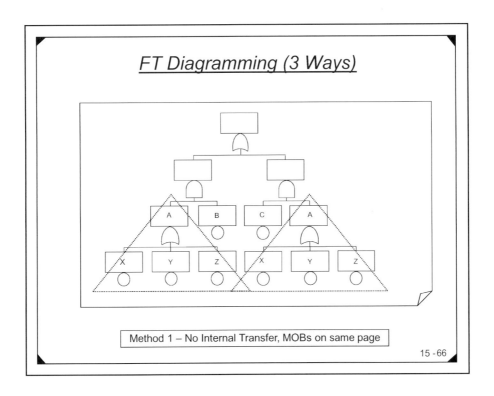

Chapter 15 – Fault Tree Analysis

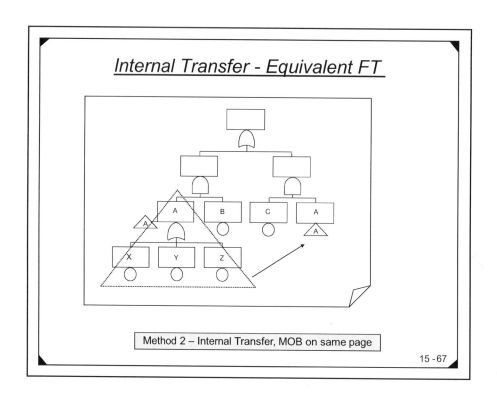

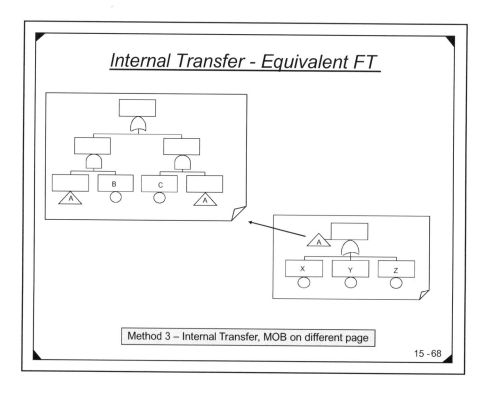

Chapter 15 – Fault Tree Analysis

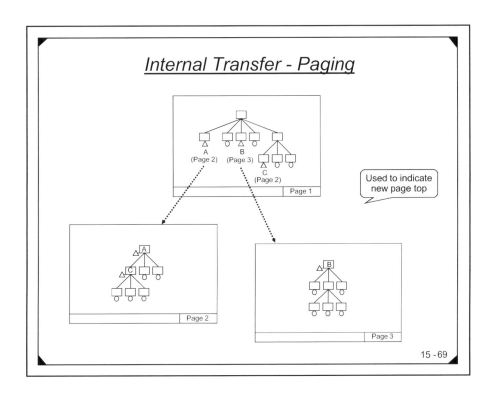

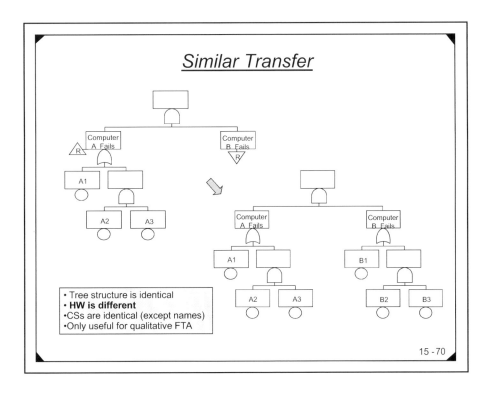

Chapter 15 – Fault Tree Analysis

External Transfer

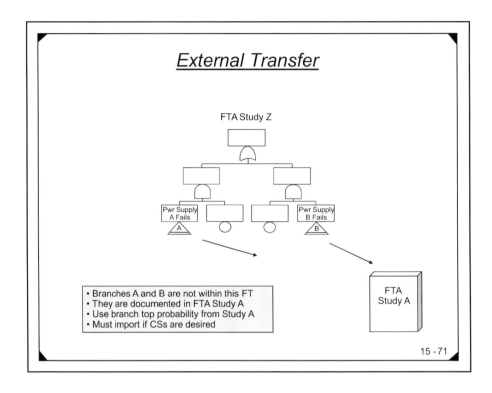

- Branches A and B are not within this FT
- They are documented in FTA Study A
- Use branch top probability from Study A
- Must import if CSs are desired

15 - 71

Failure / Fault

- Failure
    - The occurrence of a *basic component failure*.
    - The result of an internal inherent failure mechanism, thereby requiring no further breakdown.
    - Example - *Resistor R77 Fails in the Open Circuit Mode.*
- Fault
    - The occurrence or existence of an *undesired state* for a component, subsystem or system.
    - The result of a failure or chain of faults/failures; can be further broken down.
    - The component operates correctly, except at the wrong time, because it was commanded to do so.
    - Example – The light is failed off because the switch failed open, thereby removing power.

15 - 72

Chapter 15 – Fault Tree Analysis

Failure / Fault Example

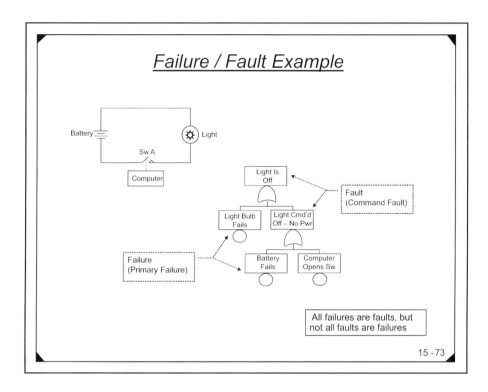

All failures are faults, but not all faults are failures

15 - 73

Independent / Dependent Failure

- Independent Failure
    - Failure <u>is not</u> caused or contributed to by another event or component

- Dependent Failure
    - Failure <u>is</u> caused or contributed to by another event or component
    - A component that is caused to fail by the failure of another component
    - The two failure are directly related, and the second failure depends on the first failure occurring
    - Example - An IC fails shorted, drawing high current, resulting in resistor R77 failing open

Dependency complicates the FT mathematics

15 - 74

Chapter 15 – Fault Tree Analysis

Primary Failure

- An inherent component failure mode
- Basic FT event
- A component failure that cannot be further defined at a lower level
- Example – diode inside a computer fails due to materiel flaw
- Symbolized by a Circle
- Has a failure rate (λ) or probability of failure

Resistor R77 fails open

Secondary Failure

- A component failure that is caused by an external force to the system
- Basic FT event
- Example – Integrated circuit fails due to external RF energy
- Important factor in Common Cause Analysis
- Symbolized by a Diamond
- Has a failure rate (λ) or probability of failure

Resistor R77 fails open from excessive RF energy

Chapter 15 – Fault Tree Analysis

Undeveloped Failure

- A component failure that can be further defined at a lower level of detail, but is not for various reasons
  - Ground rules
  - Save analysis time and money
  - May not be a critical part of FTA
- Example – computer fails (don't care about detail of why)
- Basic FT event
- Symbolized by a Diamond
- Has a failure rate (λ) or probability of failure

Computer CC107 fails to operate

15 - 77

Command Failure

- A fault state that is commanded by an upstream fault / failure
- Normal operation of a component, except in an inadvertent or untimely manner. The normal, but, undesired state of a component at a particular point in time
- The component operates correctly, except at the wrong time, because it was commanded to do so by upstream faults
- Example – a bridge opens (at an undesired time) because someone accidentally pushed the Bridge Open button
- Symbolized by a gate event requiring further development

15 - 78

Chapter 15 – Fault Tree Analysis

FT Time Parameters

- Mission Time
  - The length of time the system is in operation to complete the mission
  - Most equipment is in operation during this period of time
- Exposure Time
  - The length of time a component is effectively exposed to failure during system operation ($P = 1.0 - e^{-\lambda T}$)
  - The time assigned to equipment in FT probability calculations Exposure time can be controlled by design, repair, circumvention, testing and monitoring
- Fault Duration Time
  - The length of time a component is effectively in the failed state
  - This state is ended by repair of the component or by system failure
- Hazard Duration Time
  - The length of time a hazard exists

$$P = 1.0 - e^{-\lambda T}$$ (Time)

System Complexity Terms

- MOE
  - A Multiple Occurring Event or failure mode that occurs more than one place in the FT
  - Also known as a redundant or repeated event
- MOB
  - A multiple occurring branch (i.e., a repeated branch)
  - A tree branch that is used in more than one place in the FT
  - All of the Basic Events within the branch would actually be MOE's
- Branch
  - A subsection of the tree (subtree), similar to a limb on a real tree
- Module
  - A subtree or branch
  - An independent subtree that contains no outside MOE's or MOB's, and is not a MOB

Chapter 15 – Fault Tree Analysis

Example

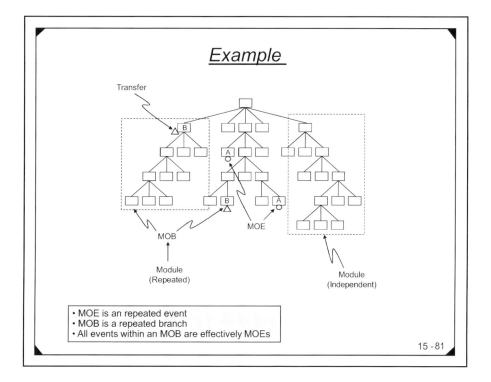

- MOE is an repeated event
- MOB is a repeated branch
- All events within an MOB are effectively MOEs

Cut Set Terms

- Cut Set
  - A set of events that together cause the tree Top UE event to occur
- Min CS (MCS)
  - A CS with the minimum number of events that can still cause the top event
- Super Set
  - A CS that contains a MCS plus additional events to cause the top UE
- Critical Path
  - The highest probability CS that drives the top UE probability
- Cut Set Order
  - The number of elements in a cut set
- Cut Set Truncation
  - Removing cut sets from consideration during the FT evaluation process
  - CS's are truncated when they exceed a specified order and/or probability

Chapter 15 – Fault Tree Analysis

Cut Set

- A unique set of events that together cause the Top UE event to occur
- One unique root cause of the Top UE (of possibly many)
- A CS can consist of one event or multiple simultaneous events or elements

Note:
A CS element can be a:
- Failure
- Human error
- Software anomaly
- Environment condition
- Normal action

The Value of Cut Sets

- CSs identify which unique event combinations can cause the UE
- CSs provide the mechanism for probability calculations
- CSs reveal the critical and weak links in a system design
  - High probability
  - Bypass of intended safety or redundancy features

Note:
Always check all CS's against the system design to make sure they are valid and correct.

Chapter 15 – Fault Tree Analysis

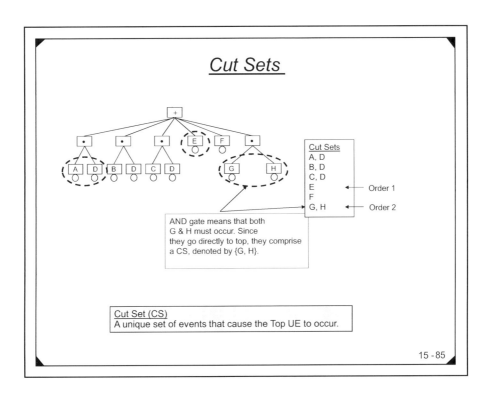

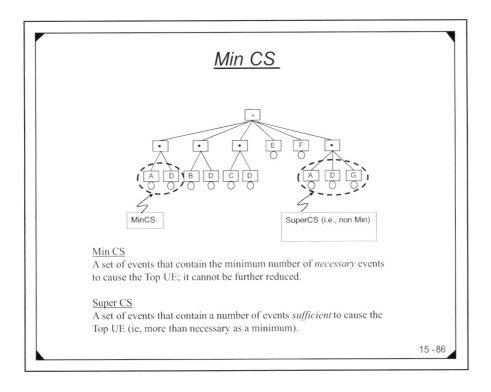

Chapter 15 – Fault Tree Analysis

Min CS Example

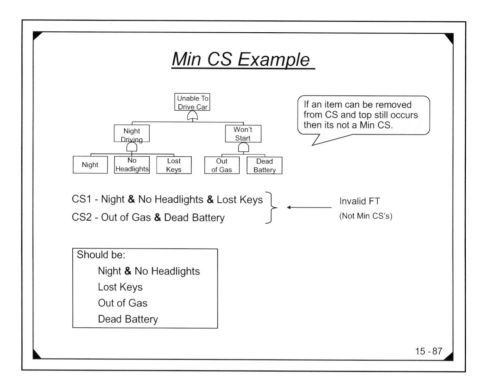

CS1 - Night **&** No Headlights **&** Lost Keys
CS2 - Out of Gas **&** Dead Battery

← Invalid FT (Not Min CS's)

Should be:
    Night **&** No Headlights
    Lost Keys
    Out of Gas
    Dead Battery

15 - 87

Min CS

- A CS with the minimum number of events that can still cause the top event
- The true list of CS's contributing to the Top
- The final CS list after removing all SCS and DupCS
- Additional CS's are often generated, beyond the MinCS's
    - Super Cut Sets (SCS) – result from MOE's
    - Duplicate Cut Sets (DupCS) - result from MOE's or AND/OR combinations
- Why eliminate SCS and DupCS?
    - Laws of Boolean algebra
    - Would make the overall tree probability slightly larger (erroneous but conservative)

15 - 88

Chapter 15 – Fault Tree Analysis

Min CS

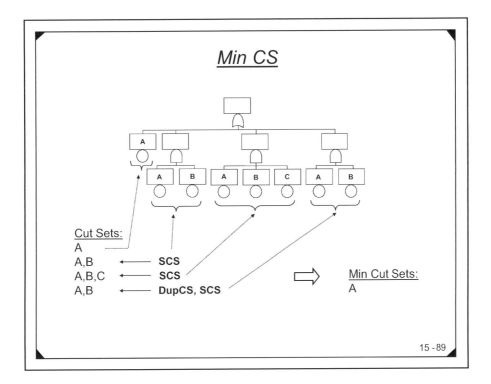

Cut Sets:
A
A,B ← **SCS**
A,B,C ← **SCS**
A,B ← **DupCS, SCS**

⇒ Min Cut Sets:
A

Importance Measure

- Importance Measure
    - A measure of the relative importance a basic event has in the tree
    - Reducing either the exposure or improving the failure rate of an Important event would make improvements in the overall tree probability
- Basic Event Importance Measure
    - A measure of the contribution a basic event has to the occurrence of the Top Event
- Basic Event Sensitivity
    - A measure of the contribution of a Basic Event to the occurrence of the Top Event relative to the Basic Events Failure rate or probability of occurrence
- Basic Event Criticality
    - Basic Event Criticality in most cases is a measure of the contribution of a Basic Event to the Top Events occurrence based upon the relative position the Basic Event holds in the tree

Chapter 15 – Fault Tree Analysis

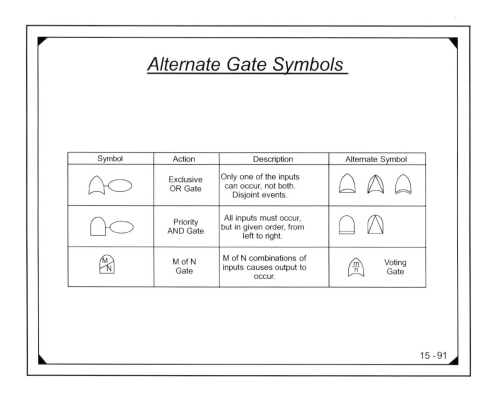

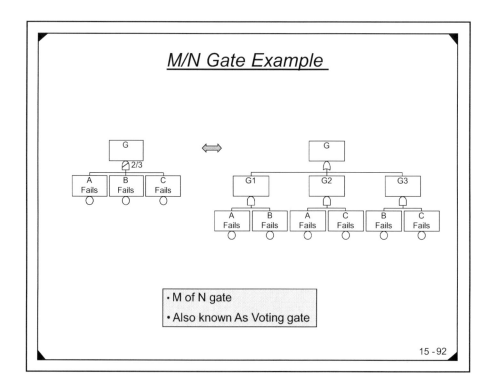

Chapter 15 – Fault Tree Analysis

--- FT Construction Process ---

- Tree is developed in:
    - Layers
    - Levels
    - Branches

- Tree Levels:
    - Top Level
        - Defines the top in terms of discrete system functions that can cause the top UE
        - Shapes the overall structure of the tree
    - Intermediate Level
        - Defines the logical relationships between system functions and component behavior
        - Function – systems – subsystems – modules - components
    - Bottom Level
        - Consists of the Basic Events or component failure modes

Overview

- Iterative process
    - Repeat the same thought process until each branch is complete
    - Use the basic construction method for each Intermediate Gate

- A branch is complete when:
    - The Basic Event components are reached
    - The previously defined boundaries are reached
    - It is recognized that there is no value in going further in detail

- Need to dig deep enough to know:
    - No critical items or functions have been missed
    - All paths have been carried far enough eliminate the possibility of common events (MOE's) in the branches

Chapter 15 – Fault Tree Analysis

FT Levels and Elements

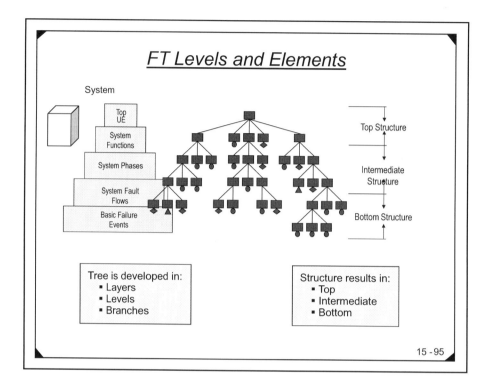

Tree is developed in:
- Layers
- Levels
- Branches

Structure results in:
- Top
- Intermediate
- Bottom

Basic FT Construction Theory

- Work down each FT branch in a repetitive (iterative) process
  - Continuously identify all Cause-Effect events and relationships
- Review and expand the event under investigation
  - Identify all the possible causes of this event (deductive evaluation)
  - Look to the next *Immediate* event(s), but not further
  - Do not jump ahead
  - Ask "what is *Necessary* and *Sufficient*"
  - Identify the relationship or logic of the Cause-Effect events – the gate type
- Structure the tree with these events and logic gate
- Keep looking back
  - To ensure identified events are not repeated
  - Do not get so engrossed in detail that the big picture is overlooked

Chapter 15 – Fault Tree Analysis

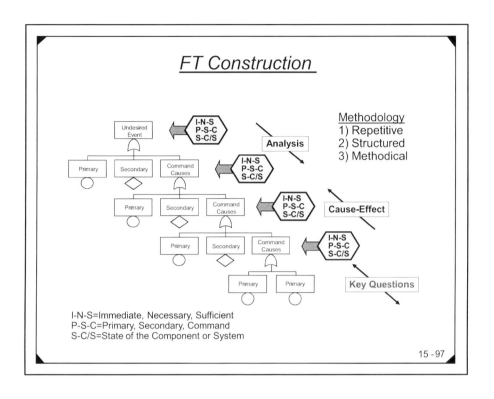

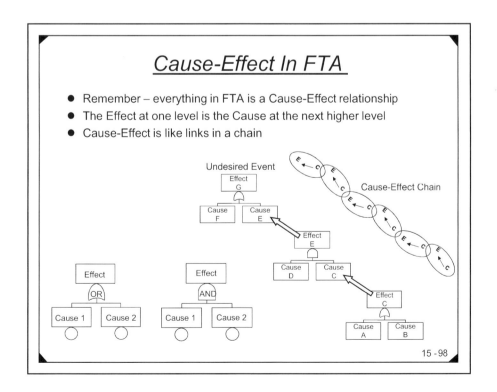

Chapter 15 – Fault Tree Analysis

FT Construction -- Iterative Process

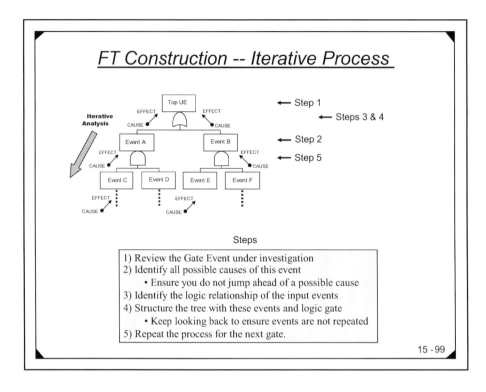

Steps
1) Review the Gate Event under investigation
2) Identify all possible causes of this event
   • Ensure you do not jump ahead of a possible cause
3) Identify the logic relationship of the input events
4) Structure the tree with these events and logic gate
   • Keep looking back to ensure events are not repeated
5) Repeat the process for the next gate.

What & How

- Ask "What" and "How"
  - What faults will cause the event?
  - How are these faults combined?
  - Always remember -- Immediate & Necessary & Sufficient
- *What* establishes the input events
- *How* establishes the Gate type
  - AND gate – all inputs are required
  - OR gate – any one or all inputs are required

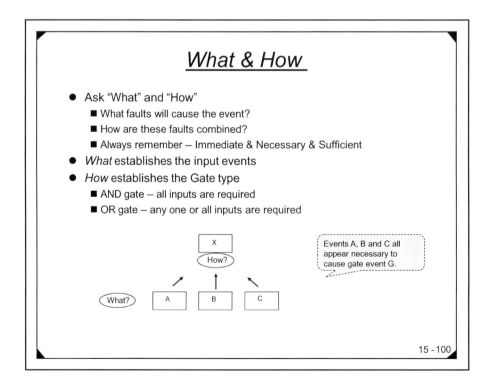

Events A, B and C all appear necessary to cause gate event G.

Chapter 15 – Fault Tree Analysis

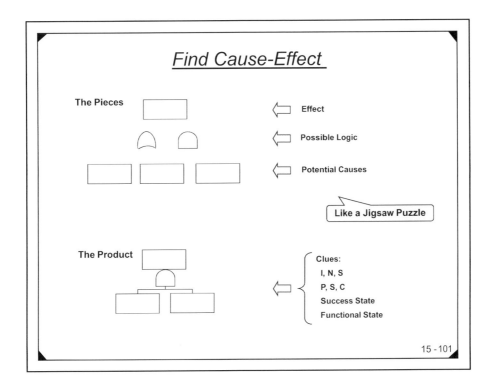

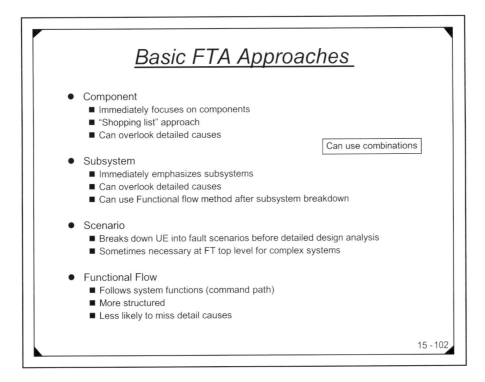

Chapter 15 – Fault Tree Analysis

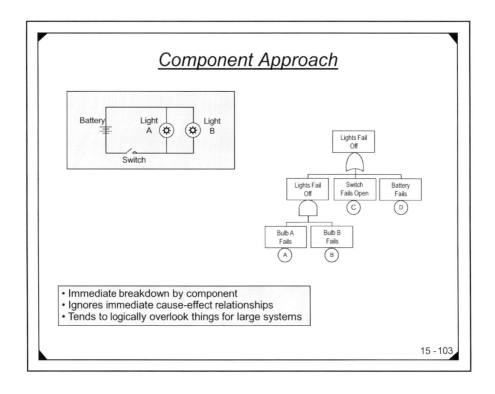

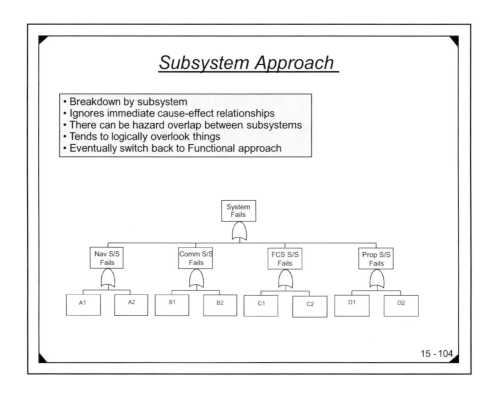

Chapter 15 – Fault Tree Analysis

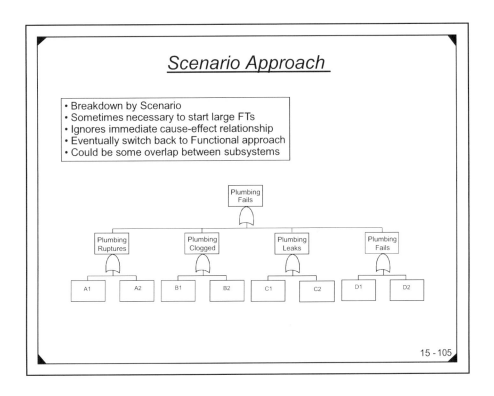

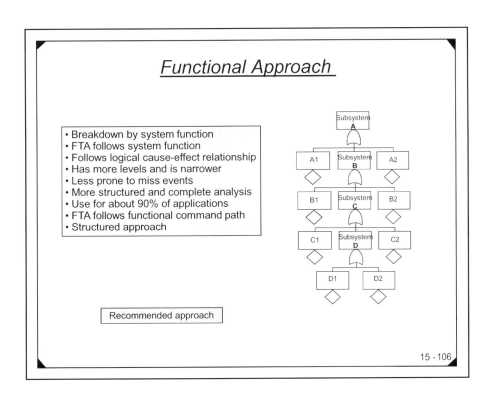

Chapter 15 – Fault Tree Analysis

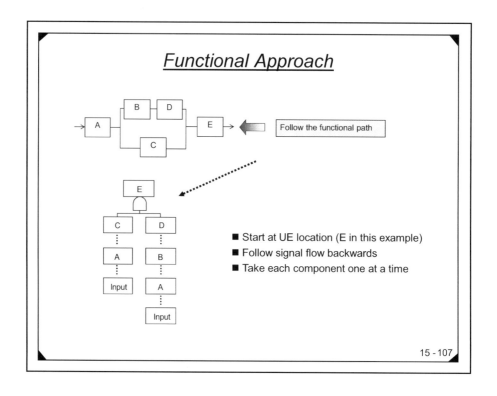

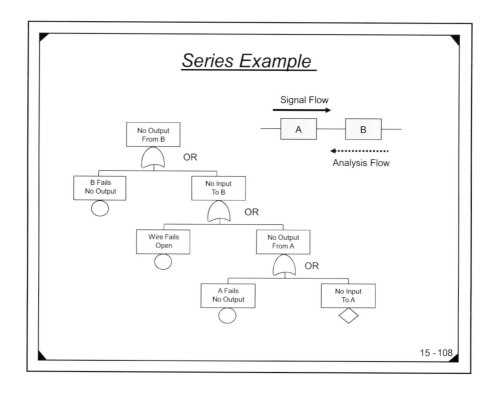

Chapter 15 – Fault Tree Analysis

Series-Parallel Example

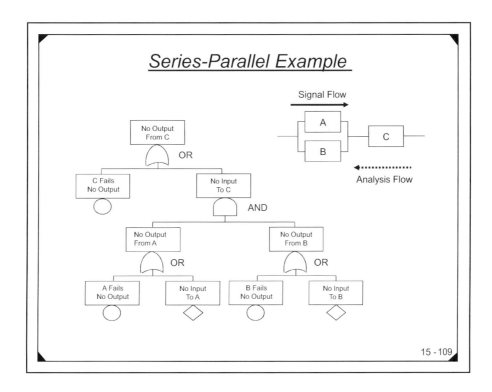

Gate Node Development

- Construction at each gate involves a 3 step question process:
  - Step 1 – Immediate, Necessary and Sufficient (I-N-S) ?
  - Step 2 – Primary, Secondary and Command (P-S-C) ?
  - Step 3 – State of the Component or System (S-C/S) ?

These are the 3 key questions in FTA construction

Step 1

- Step 1 – *What is* Immediate, Necessary and Sufficient (I-N-S) ?
    - Read the gate event wording
    - Identify all **Immediate**, **Necessary** and **Sufficient** events to cause the Gate event
        - Immediate – do not skip past events
        - Necessary – include only what is actually necessary
        - Sufficient – do not include more than the minimum necessary
    - Structure the I-N-S casual events with appropriate logic
    - Mentally test the events and logic until satisfied

15 - 111

Step 1

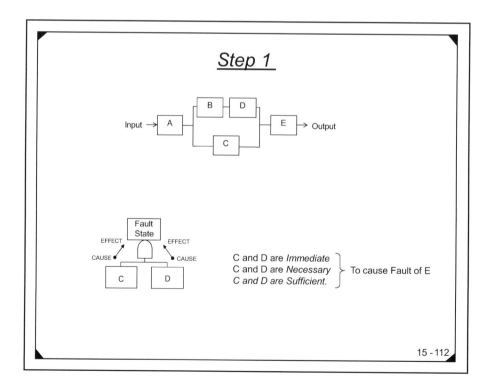

C and D are *Immediate*
C and D are *Necessary* } To cause Fault of E
C and D are *Sufficient.*

15 - 112

Chapter 15 – Fault Tree Analysis

Step 2

- Step 2 – *What is* Primary, Secondary and Command (P-S-C) ?
  - Read the gate event wording
  - Review I-N-S events from Step 1
  - Identify all **Primary**, **Secondary** and **Command** events causing the Gate event
    - Primary Fault – basic inherent component failure
    - Secondary Fault – failure caused by an external force
    - Command Fault – A fault state that is commanded by an upstream fault or failure
  - Structure the P-S-C casual events with appropriate logic

> If there are P-S-C inputs, then it's an OR gate

Step 2

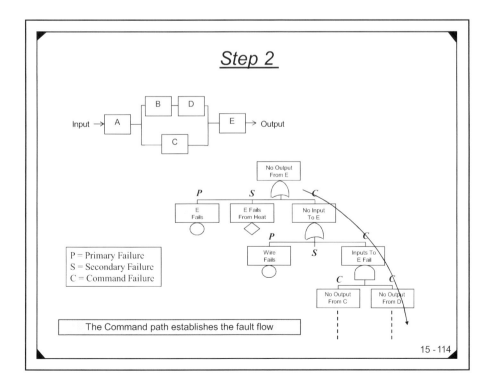

P = Primary Failure
S = Secondary Failure
C = Command Failure

The Command path establishes the fault flow

Chapter 15 – Fault Tree Analysis

Step 3

- Step 3 – *Is it a* State of the Component or System (S-C/S) *fault* ?
    - Read the gate event wording
    - Identify if the Gate involves
        - a *State of the Component* fault
            - **Being directly at the component level**
            - **Evaluating the causes of a component failure**
        - a *State of the System* fault
            - **Being a system level event**
            - **If it's not a state of the component fault**
    - Structure the casual events with appropriate logic

Step 3 (continued)

- If State of the Component, then:
    - Ask "what are the P-S-C causes"
    - Generally this results in an OR gate
    - If a Command event is not involved, then this branch path is complete

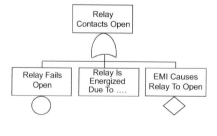

Chapter 15 – Fault Tree Analysis

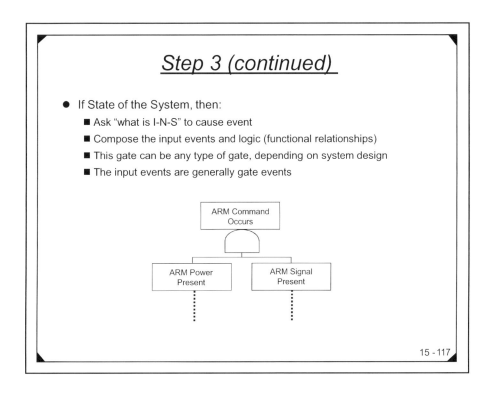

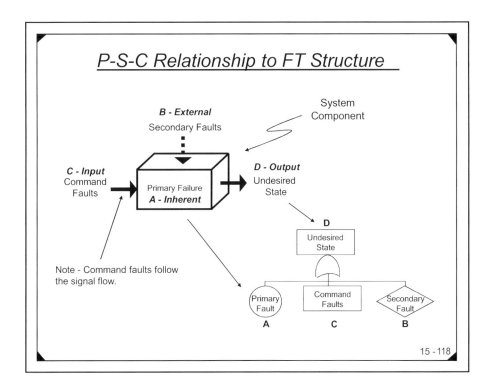

Chapter 15 – Fault Tree Analysis

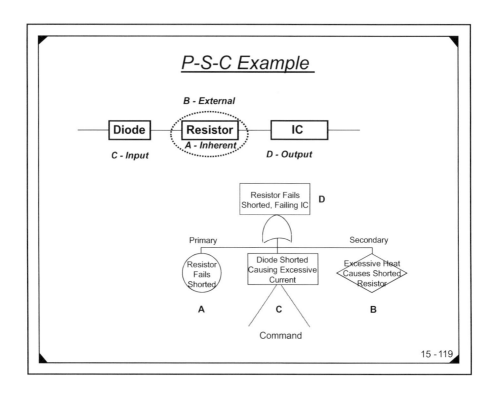

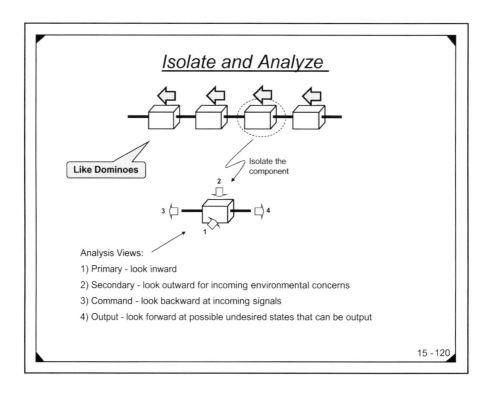

Chapter 15 – Fault Tree Analysis

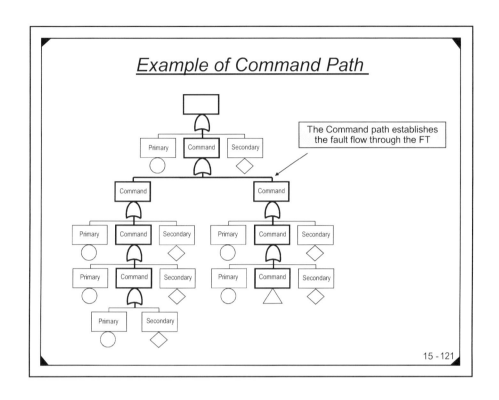

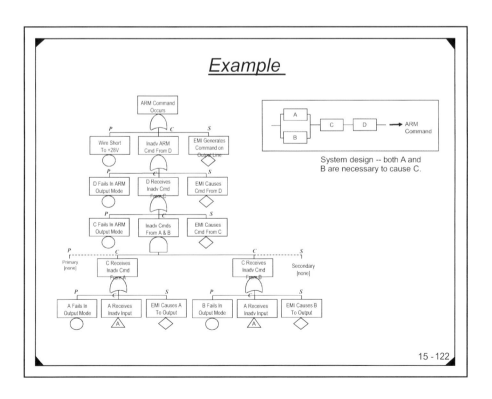

Chapter 15 – Fault Tree Analysis

Example Fault Tree

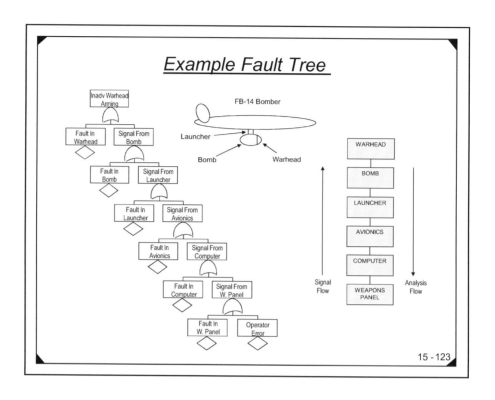

15 - 123

System Component Types

- Active -- contribute in a dynamic manner ← *Initiator*
  - Provides or controls the force, power, energy, etc. that does the work required for system operation
  - e.g., relays, valves, electrical power
  - Reliability may be a concern, thus requiring design redundancy

- Passive -- contribute in a static manner ← *Enabler*
  - Provide the path for the energy source
  - e.g., wires, pipes
  - Typically passive components are much more reliable (SPFs can be more easily tolerated)

15 - 124

Chapter 15 – Fault Tree Analysis

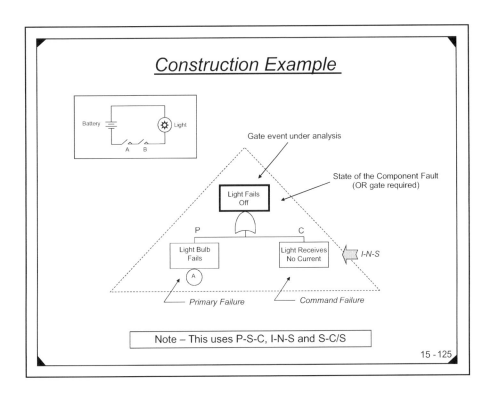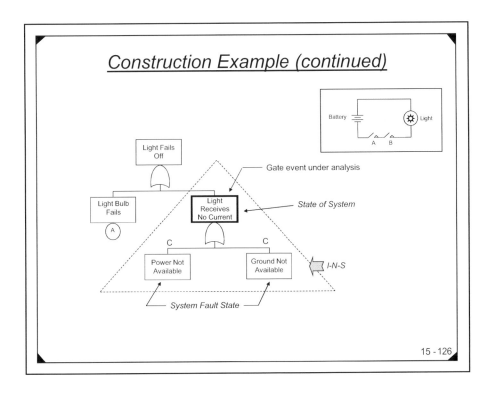

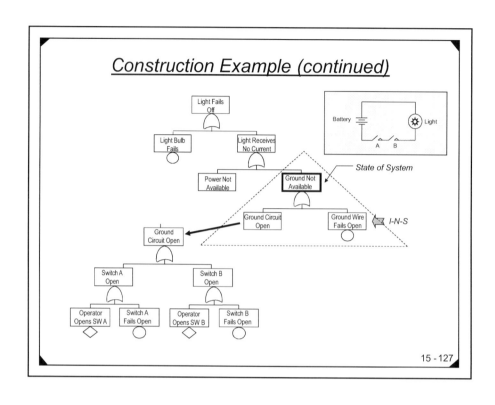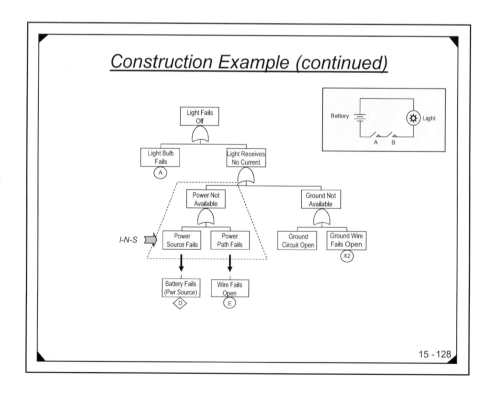

Chapter 15 – Fault Tree Analysis

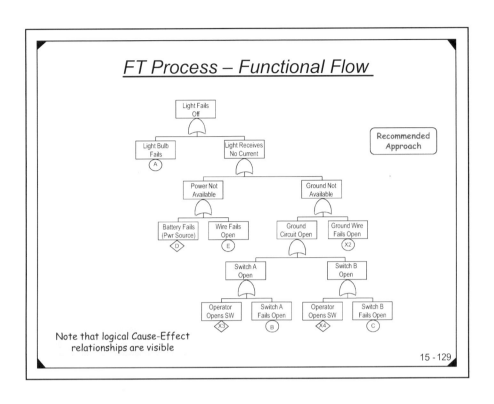

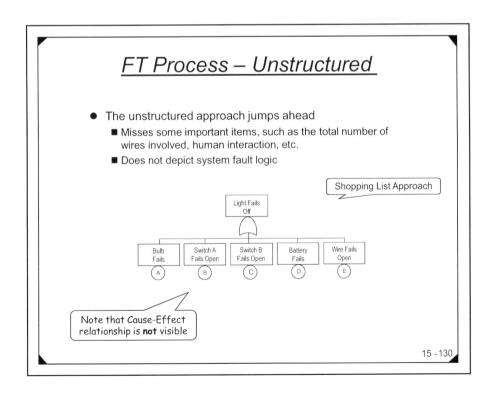

Chapter 15 – Fault Tree Analysis

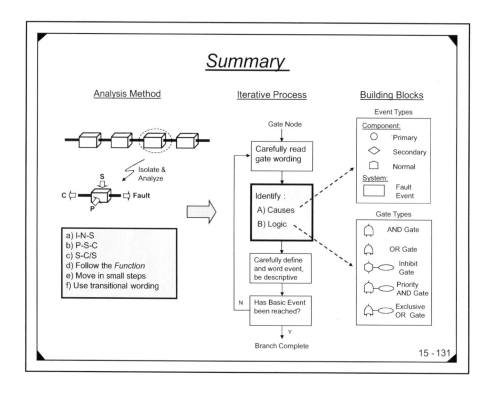

--- FT Construction Rules ---
1 - Know Your System

- It is imperative to know and understand the system design and operation thoroughly
- Utilize all sources of design information
    - Drawings, procedures, block diagrams, flow diagrams, FMEAs
    - Stress analyses, failure reports, maintenance procedures
    - System interface documents
    - CONOPS
- Drawings and data must be current for current results
- Draw a Functional Diagram of the system

Rule of thumb - if you can't construct a block diagram of system you may not understand it well enough to FT

Chapter 15 – Fault Tree Analysis

2 - Understand The Purpose Of Your FTA

- It's important to know why the FTA is being performed
  - To ensure adequate resources are applied
  - To ensure proper scope of analysis
  - To ensure the appropriate results are obtained

- Remember, FTA is a tool for
  - Root cause analysis
  - Identifies events contributing to an Undesired Event
  - Computes the probability of an Undesired Event
  - Measures the relative impact of a design fix
  - Logic diagrams for presentation

3 - Understand Your FT Size

- FT size impacts the entire FTA process
- As FTs grow in size many factors are affected
  - Cost (e.g., manpower)
  - Time
  - Complexity
  - Understanding
  - Traceability
  - Computation
- System factors that cause FT growth
  - System size
  - Safety criticality of system
  - System complexity
- FT factors that cause FT growth
  - MOEs and MOBs (e.g., redundancy)
  - Certain AND / OR combinations

FT size is important and has many implications

Chapter 15 – Fault Tree Analysis

4 - Intentionally Design Your Fault Tree

- As a FT grows in size it is important develop an architecture and a set of rules
- The architecture lays out the overall FT design
  - Subsystem branches (for analysts and subcontractors)
  - Analyst responsibilities
- The rules provide consistent development guidelines
  - Ground rules for inclusion/exclusion (e.g., Human factors, CCFs)
  - Ground rules for depth of analysis (subsystem, LRU, component)
  - Ground rules for naming convention
  - Ground rules for component database

Foresight helps avoid future problems

Don't Do This! -- Plan Ahead

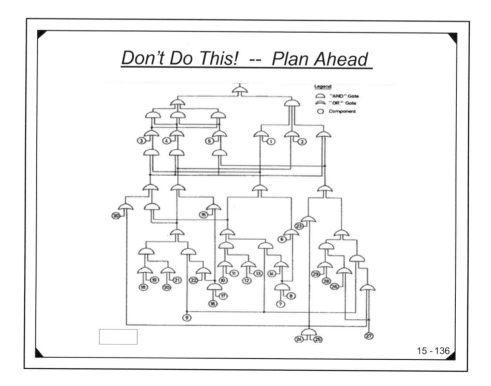

Chapter 15 – Fault Tree Analysis

FT By Design

- FTA projects require planning, organization, structure, foresight -- *design*
  - Due to tree size, complexity and FT team communications
- Proper FT design or architecture can prevent:
  - Major rework for small changes
  - Errors in tree structure, tree data, evaluation
- Considerations include:
  - Allow for future expansions in tree
  - Allow for ease of modification
  - Establish name conventions for accounting and control
    - Events, transfers, MOE's, subtrees

5 - Ensure the FT is Correct and Complete

- FT completeness is critical
  - Anything left out of the FTA skews the answer
  - The final result will only reflect what was included in the FT
  - The FTA is not complete until all root causes have been identified
- FT correctness is critical
  - If the FT is not correct the results will not be accurate
- Conduct FT peer review to ensure completeness/correctness
  - Involve other FT experts
  - Involve system designers
- Items often overlooked in FTA
  - Human error
  - Common cause failures
  - Software factors (design may have dependencies)
  - Components or subsystems considered not applicable

FT results are skewed if the FT is not complete and correct

Chapter 15 – Fault Tree Analysis

6 - Know Your Fault Tree Tools

- Know basic FT tool capabilities
  - Construction, editing, plotting, reports, cut set evaluation
- Know FT tool user friendliness
  - Intuitive operation
  - Easy to use and remember
  - Changes are easy to implement
- Single vs. multi-phase FT
- Qualitative vs. quantitative evaluation
- Simulation vs. analytical evaluation (considerations include size, accuracy, phasing)

Tools (continued)

- Know FT tool limitations
  - Tree size (i.e., max number of events)
  - Cut set size
  - Plot size
- Understand approximations and cutoff methods, some can cause errors
- Gate probabilities could be incorrect when MOEs are involved
- Test the tool; don't assume answers are always correct

Don't place complete trust in a FT program

Chapter 15 – Fault Tree Analysis

7 - Understand Your FTA Results

- Verify that the FTA goals were achieved
    - Was the analysis objective achieved
    - Are the results meaningful
    - Was FTA the right tool
    - Are adjustments necessary

- Make reasonableness tests to verify the results
    - Are the results correct
    - Look for analysis errors (logic, data, model, computer results)
    - Are CSs credible and relevant (if not revise tree)
    - Take nothing for granted from the computer
    - Test your results via manual calculations

8 - Document Your FTA

- Formally document the entire FTA
    - May need to provide to customer (product)
    - May need to defend at a later date
    - May need to modify at a later date
    - May perform a similar analysis at a later date
    - May need records for an accident/incident investigation

- Even a small analysis should be documented for posterity

- May support future questions or analyses

Documentation is essential

Chapter 15 – Fault Tree Analysis

Documentation (continued)

- Provide complete documentation
  - Problem statement
  - Definitions
  - Ground rules
  - References
  - Comprehensive system description
  - Data and sources (drawings, failure rates, etc.)
  - FT diagrams
  - FT tree metrics
  - FT computer tool description
  - Results
  - Conclusions

> Document the number of hours to perform the FTA for future estimates

10 - Think in Terms of Failure Space

- Remember, it's a "fault" tree, not a "success" tree
  - Analysis of failures, faults, errors and bad designs

- No magic
  - Do not draw the fault tree assuming the system can be saved by a miraculous failure
  - This is normally referred to as the "No Magic Rule"

- No operator *saves*
  - When constructing FT logic do not assume that operator action will save the system from fault conditions
  - Only built-in safety features can be considered
  - Operator errors can be considered in the FT, but not operator saves
  - The system design is under investigation, not the operator performing miracles

Chapter 15 – Fault Tree Analysis

11 - Correct Node Wording Is Important

- Be clear and precise
- Express fault event in terms of
  - Device transition
  - Input or output state
- Be very descriptive in writing event text
  - "Power supply fails" vs. "Power supply does not provide +5 VDC"
  - "Valve fails in closed position" vs. "Valve fails"
- Do not
  - Use the terms Primary, Secondary or Command
    - Thought process
    - Symbols already show it
  - Use terms Failure or Fault (if possible) – not enough information

Good node wording guides the analysis process

15 - 145

Wording Example

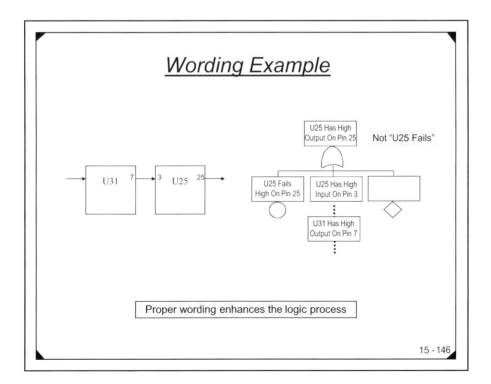

Not "U25 Fails"

Proper wording enhances the logic process

15 - 146

Chapter 15 – Fault Tree Analysis

12 - Follow Standard Construction Rules

- No gate-to-gate diagrams
  - Do not draw a gate without a gate node box and associated descriptive text and rectangle

- Use only one output from a node
  - Do not connect the output of a node to more than one input nodes.
  - Some analysts attempt to show redundancy this way, but it becomes cluttered and confusing.
  - Most computer codes cannot handle this situation anyway.

Construction Errors

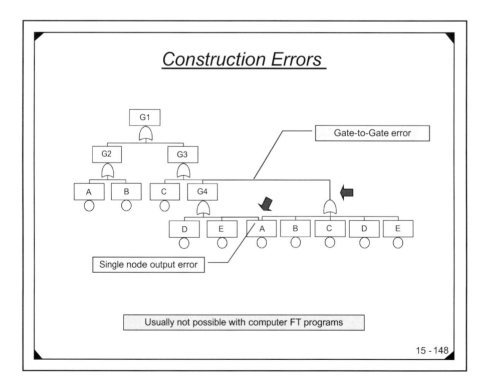

Usually not possible with computer FT programs

Chapter 15 – Fault Tree Analysis

FT Construction Rules (cont'd)

- Construct the FT to most accurately reflect the system design and logic
  - Do not try to modify the tree structure to resolve an MOE.
  - Let the FT computer software handle all MOE resolutions.

- Keep single input OR gates to a minimum
  - When the words in a Node box exceed the box limit, you can create another input with a Node box directly below just to continue the words
  - Use the Notes if additional words are needed. Its okay to do but prudence is also necessary

- Use House events carefully
  - A House (Normal event) never goes into an OR gate, except in special cases, such as a multi-phase simulation FT

FT Construction Rules (cont'd)

- Do not label fault events on the tree as *Primary*, *Secondary* and *Command* failures
  - Go into detail and be descriptive. These terms are more for the thought process than the labeling process.

- When possible add traceability detail
  - Put drawing numbers and part numbers in the fault event or in the notes.
  - This provides better traceability when the tree is being reviewed or checked, or when the tree is being modified after a lengthy time period.

Chapter 15 – Fault Tree Analysis

FT Construction Rules (cont'd)

- Operator error should be included in the analysis where appropriate
  - It is up to the analyst and the purpose/objective of the FTA as to whether the event should be included in quantitative evaluations
  - The decision needs to be documented in the analysis ground rules

- Take a second look at all tree logic structure
  - Sometimes what appears to be a simple and correct tree logic structure might actually be flawed for various reasons
    - Example -- mutually exclusive events, logic loops, etc.
  - Make sure there are no leaps or gaps in logic
  - The tree structure may need revising in these cases

13 - Provide Necessary Node Data

- Node name
- Node text
- Node type
- Basic event probability (for quantification only)

Four items are essential

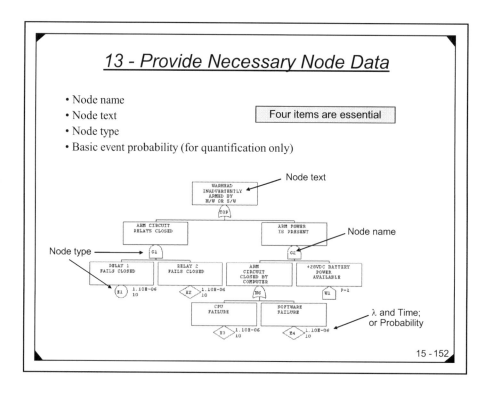

Chapter 15 – Fault Tree Analysis

14 - Apply FT Aesthetics

- When the FT structure looks good it will be better accepted
- A level FT structure looks best
  - No zig-zags
- Balance page breaks & FT structure
  - Avoid too little info on a page (i.e., 2 or 3 events)
- Always use standard FT symbols (defined in NUREG book)
- Computerized construction tools provides better graphics than manual methods

A level and balanced FT structure is easier to read

Poor Aesthetics Example

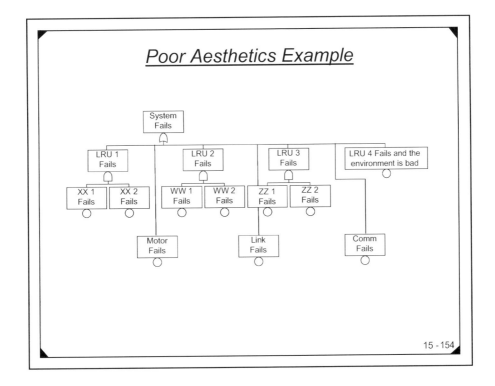

Chapter 15 – Fault Tree Analysis

15 - Computerized Evaluation Is Essential

- FT quantification is easy when the FT is small and simple
  - Manual calculations are easy

- FT quantification is difficult when the FT is large and complex
  - Manual quantification becomes too difficult without errors

16 - Validate all CSs

- CSs are very important
  - They show where to fix system (weak design points)
  - They show the importance of specific components
  - They are necessary for most numerical calculations

- Always verify that all CSs are valid
  - If they are not right the FT is incorrect

Chapter 15 – Fault Tree Analysis

17 - Perform a Numerical Reality Check

- Never completely trust the results of a computer program
  - Some algorithms may have errors
  - Proprietary approximations may not always work

- Perform a rough calculation manually to check on the computer results

- A large deviation could indicate a problem

18 - Verify All MOEs and MOBs

- Review MOEs very carefully
  - Their effect can be important - common cause, zonal analysis
  - They can cause large numerical error (or none at all)
  - They can hide or emphasize redundancy

- An MOE or MOB can be inadvertently created by erroneously using the same event name twice

Chapter 15 – Fault Tree Analysis

19 - FTs Are Only Models

- Remember that FT's are models
    - Perception or model of reality
    - Not 100% fidelity to exact truth
- Remember that models are approximations (generally)
    - Not necessarily 100% exact
    - Still a valuable predictor
    - Newton's law of gravity is an approximation
- Do not represent FTA results as an exact answer
    - Use engineering judgment
    - Small number are relative (2.0×10^{-8} is as good as 1.742135×10^{-8})
    - Anything overlooked by the FTA skews the answer
        - Minor things left out can make results conservative (understate results)
        - Major things left out can be significant (overstate results)

22 - Understand Your Failure Data

- Failure data must be obtainable for quantitative evaluation
- Must understand failure modes, failure mechanisms and failure rates
- Data accuracy and trustworthiness must be known (confidence)
- Proven data is best
- Don't be afraid of raw data
    - Data estimates can be used
    - Useful for rough estimate
    - Results must be understood

Even raw data provides useful results

Chapter 15 – Fault Tree Analysis

23 – Always Provide Data Sources

- MIL-HDBK-217 Electronic Parts Predictions
- Maintenance records
- Vendor data
- Testing
- Historical databases

24 – The Human Is A System Element

- The human is often a key element in the system lifecycle
    - Manufacturing, assembly, installation, <u>operation</u>, decommissioning
- The human might be the most complex system element
- Human error includes
    - Fails to perform function (error of omission)
    - Performs incorrectly
    - Performs inadvertently (error of commission)
    - Performs wrong function
- Human error can
    - Initiate a system failure or accident
    - Fail to correctly mitigate the effects of a failure (e.g., ignored warning lights)
    - Exacerbate the effects of a system failure

Chapter 15 – Fault Tree Analysis

Include Human Error in FTs

- Human error should be considered in FT model when appropriate
    - When the probability could make a difference
    - When the design needs to be modified

- Key rule – anything left out of the FT causes the results to be understated

- A poor HSI design can force the operator to commit errors
    - Mode confusion (e.g., Predator mishap)
    - Display confusion
    - Too many screens, modes and/or functions
    - GUI Widget confusion
    - Designing the system to complement the human operator

Human Reliability is Complex

- Finding human error failure data is difficult
- Rates could theoretically vary based on many factors
    - System type
    - Design
    - Human skills
    - Repetitiveness
- In general, studies show:
    - $P = 10^{-3}$ for general error
    - $P = 10^{-4}$ to 10^{-6} if special designs and checks are performed

Chapter 15 – Fault Tree Analysis

25 - Node Name Length

- Short node names tend to be better than long names
  - Long names become burdensome & time consuming

- A 5 char name is easier to work with than a 24 char name
  - Typing original
  - Typing in a search
  - Storing in a database

- Random node names generated from node text tends to be more difficult to follow than shorted coded names

Node Naming Convention

- FT naming conventions (or coding) can be very useful
- Must maintain explicit configuration control of Event, Transfer and Gate names
  - Incorrect Event names will cause inadvertent MOE's or none when intended
  - Incorrect Transfers names will cause use of wrong modules
  - Incorrect Gate names will cause inadvertent MOB's or none when intended
- Most important for very large trees, not as critical for small trees
- Example: two analysts may use same diode, but each give it a different FT name
- A FT name coding scheme should be developed for the FT project
  - before the FT construction begins, planned, consistent

Chapter 15 – Fault Tree Analysis

Sample Coding Scheme

- Use only 5 characters for a node name
- Specific characters are used to quickly identify event types
- Establish a pattern for *tree families*

1st Char	Symbol Type
G	Gate
X	Circle (primary failure)
Z	Diamond (secondary failure)
W	House (normal event)
Y	Oval (condition event)

Chars	Family	Represents
A	Top level family	Computer System
B	Top level family	Navigation System
AA	Member of A	
BB	Member of B	
ABC	Member of AB	

Coding Example

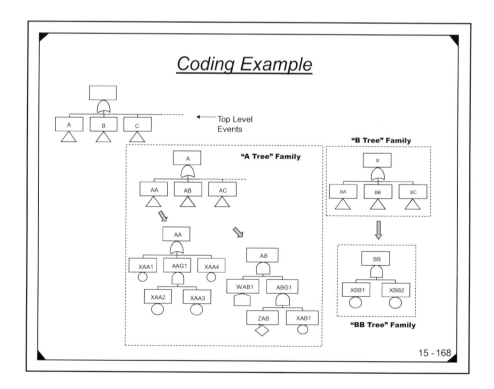

Chapter 15 – Fault Tree Analysis

26 - FT Accounting

- Large FT's necessitate FT accounting
- This is a form of data control
- Used in conjunction with the input roadmap
- Keep accurate track of basic events:
  - Name
  - Text
  - Failure rate
  - Exposure time
  - Source of data for event failure rate
  - Trees where the event is used
  - If it is an MOE
- Generally requires a database

Summary

- There are good FTs, mediocre FTs and bad FTs
- Strive to construct good FTs
- Understanding FT rules and intentionally designing a FT helps to make a quality product
- Don't just spit out a mediocre FT in order to meet a deadline or CDRL
- It's easy to visually inspect the quality of a FT
- A good FT shows credibility
- A good FT analyst can tell how much effort and <u>credibility</u> someone put into their FT just through a visual inspection
- Apply FTA rules and guidelines

Chapter 15 – Fault Tree Analysis

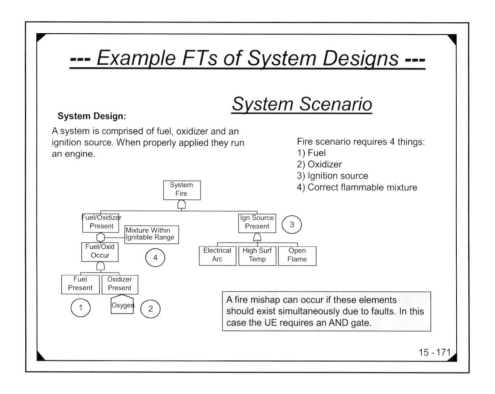

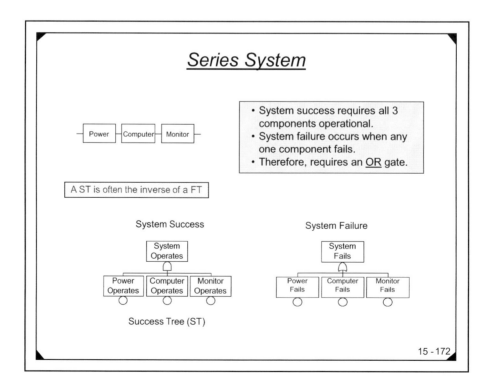

Chapter 15 – Fault Tree Analysis

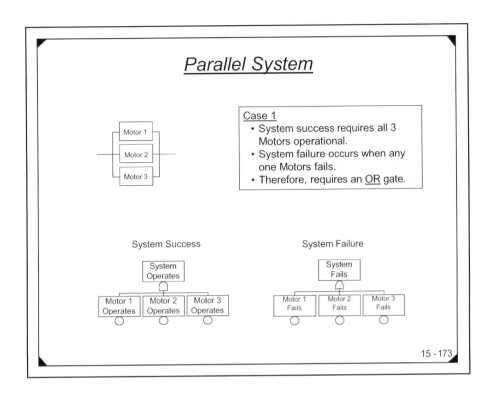

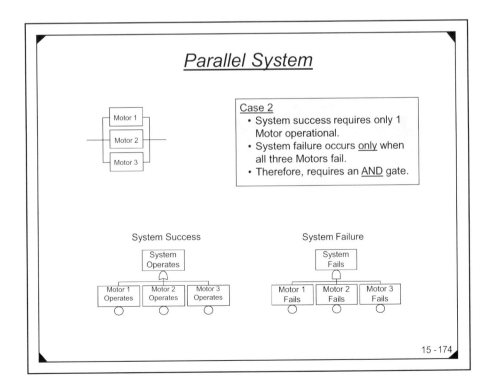

Chapter 15 – Fault Tree Analysis

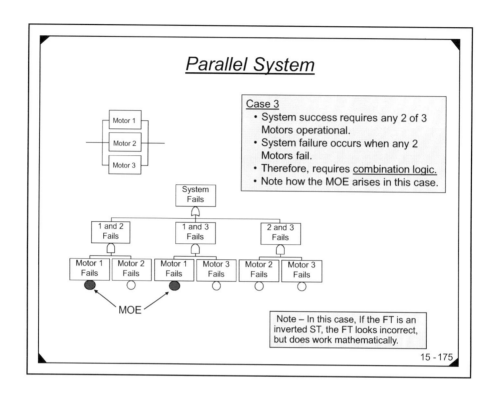

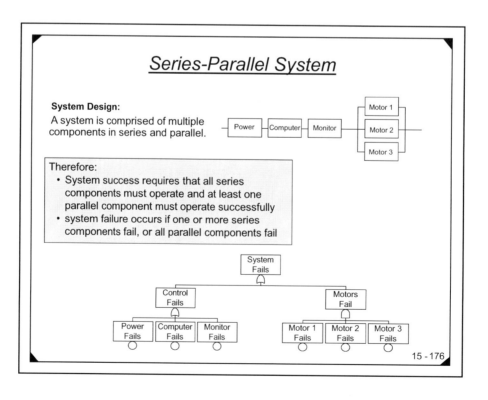

Chapter 15 – Fault Tree Analysis

Sequence Parallel System

System Design:
A system is comprised of two components A and B. System success requires that both must operate successfully at the same time. System failure occurs if both fail, but <u>only if A fails before B</u>.

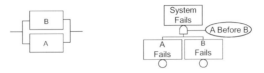

Therefore:
- Both must fail
- Sequential problem (A before B)
- The fault state logic requires a <u>Priority AND</u> gate.

Monitor System

System Design:
A system is comprised of two components, Monitor A and component B. Monitor A monitors the operation of B. If it detects any failure in B it takes corrective action. System success requires that B must operate successfully. System failure occurs if component B fails, which can only happen if Monitor A fails to detect a problem with B, and B subsequently fails. If A works it always corrects any failure in B or provides a warning.

This design has 2 different cases:
1. Full Monitor (full coverage)
2. Partial Monitor (partial coverage)

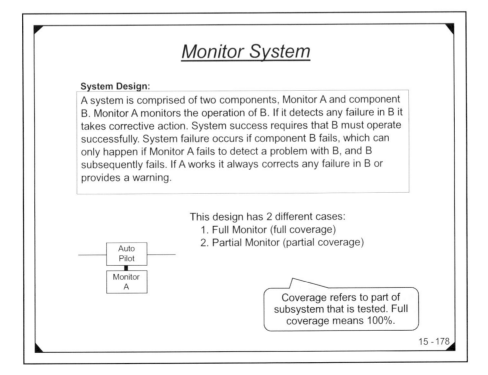

Coverage refers to part of subsystem that is tested. Full coverage means 100%.

Chapter 15 – Fault Tree Analysis

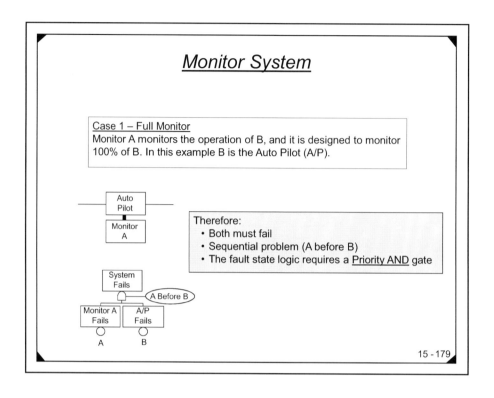

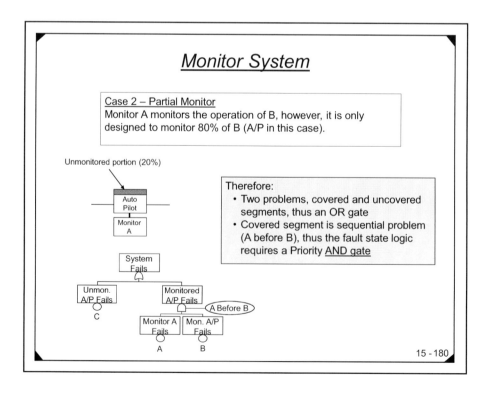

Chapter 15 – Fault Tree Analysis

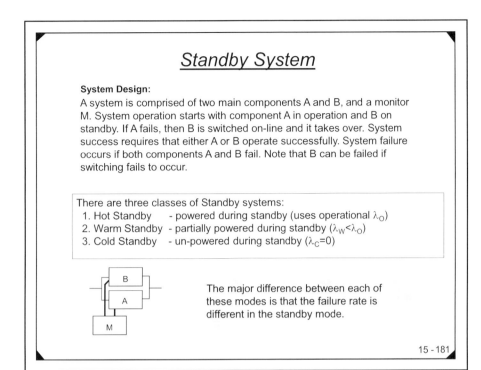

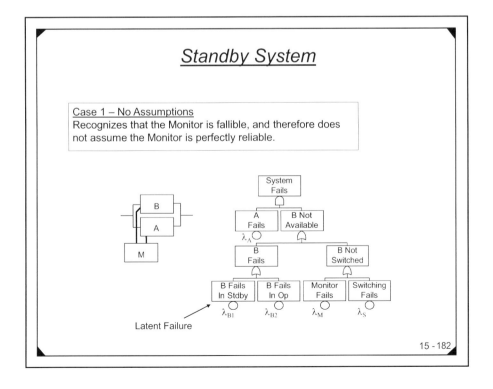

Chapter 15 – Fault Tree Analysis

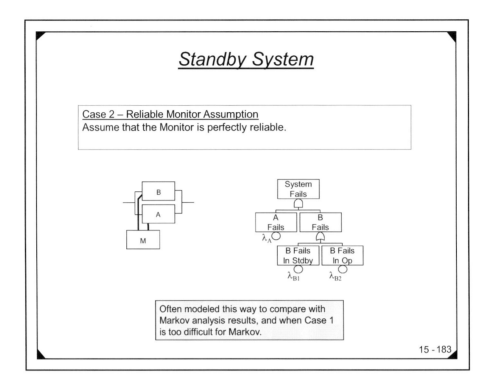

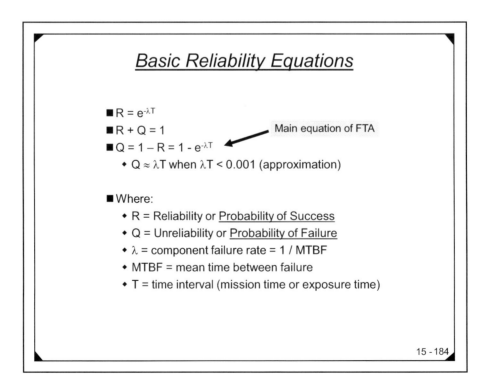

Chapter 15 – Fault Tree Analysis

Effects of Failure Rate & Time

- The longer the mission (or exposure time) the higher the probability of failure

- The smaller the failure rate the lower the probability of failure

The Effect of Exposure Time on Probability is Significant

15 - 185

FT Quantification

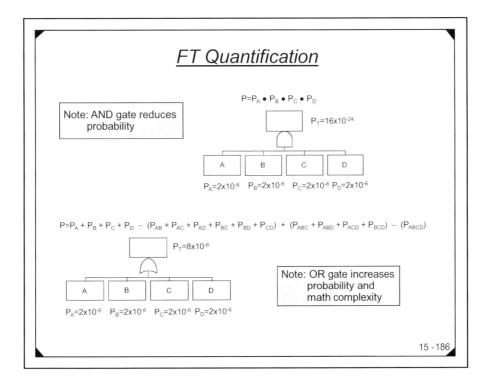

Note: AND gate reduces probability

$P = P_A \bullet P_B \bullet P_C \bullet P_D$

$P_T = 16 \times 10^{-24}$

$P_A = 2 \times 10^{-6}$ $P_B = 2 \times 10^{-6}$ $P_C = 2 \times 10^{-6}$ $P_D = 2 \times 10^{-6}$

$P = P_A + P_B + P_C + P_D - (P_{AB} + P_{AC} + P_{AD} + P_{BC} + P_{BD} + P_{CD}) + (P_{ABC} + P_{ABD} + P_{ACD} + P_{BCD}) - (P_{ABCD})$

$P_T = 8 \times 10^{-6}$

$P_A = 2 \times 10^{-6}$ $P_B = 2 \times 10^{-6}$ $P_C = 2 \times 10^{-6}$ $P_D = 2 \times 10^{-6}$

Note: OR gate increases probability and math complexity

15 - 186

Chapter 15 – Fault Tree Analysis

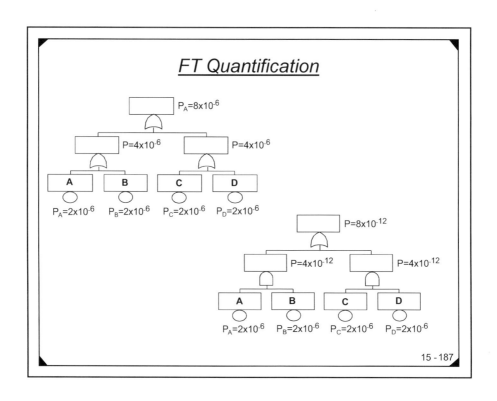

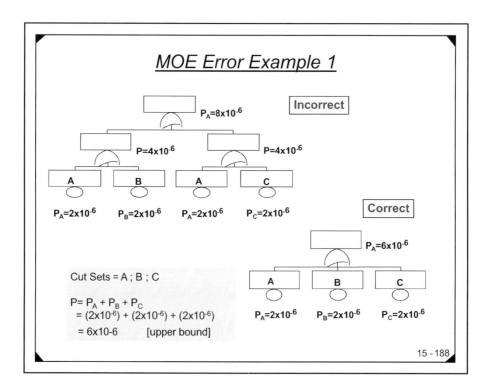

Chapter 15 – Fault Tree Analysis

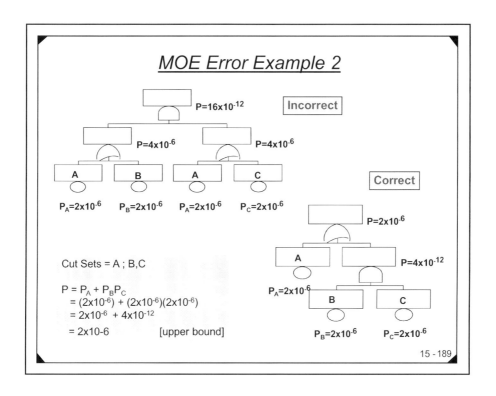

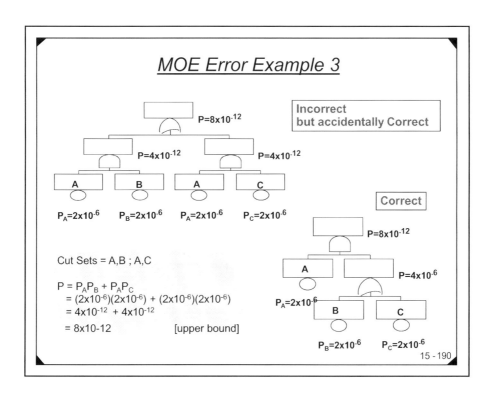

Chapter 15 – Fault Tree Analysis

CS Expansion Formula

$$P = \Sigma(\text{singles}) - \Sigma(\text{pairs}) + \Sigma(\text{triples}) - \Sigma(\text{fours}) + \Sigma(\text{fives}) - \Sigma(\text{sixes}) + \cdots$$

CS $\{A; B; C; D\}$

$$\begin{aligned}
P = &\ (P_A + P_B + P_C + P_D) \\
  & - (P_{AB} + P_{AC} + P_{AD} + P_{BC} + P_{BD} + P_{CD}) \\
  & + (P_{ABC} + P_{ABD} + P_{ACD} + P_{BCD}) \\
  & - (P_{ABCD})
\end{aligned}$$

$P = P_A + P_B + P_C + P_D - (P_{AB} + P_{AC} + P_{AD} + P_{BC} + P_{BD} + P_{CD}) + (P_{ABC} + P_{ABD} + P_{ACD} + P_{BCD}) - (P_{ABCD})$

> Size and complexity of the formula depends on the total number of cut sets and MOE's.

Min Cut Set Upper Bound Approximation

$$P = 1 - [(1 - P_{CS1})(1 - P_{CS2})(1 - P_{CS3})\ldots(1 - P_{CSN})]$$

CS$\{A; B\}$

$$\begin{aligned}
P &= 1 - [(1 - P_A)(1 - P_B)] \\
  &= 1 - [1 - P_B - P_A + P_A P_B] \\
  &= 1 - 1 + P_B + P_A - P_A P_B \\
  &= P_A + P_B - P_A P_B \quad \leftarrow \text{Equivalent to standard expansion}
\end{aligned}$$

Chapter 15 – Fault Tree Analysis

Summary

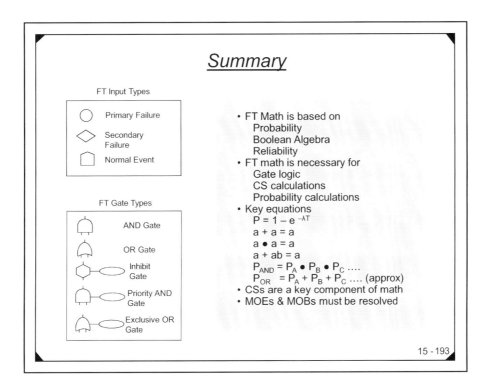

- FT Math is based on
  Probability
  Boolean Algebra
  Reliability
- FT math is necessary for
  Gate logic
  CS calculations
  Probability calculations
- Key equations
  $P = 1 - e^{-\lambda T}$
  $a + a = a$
  $a \bullet a = a$
  $a + ab = a$
  $P_{AND} = P_A \bullet P_B \bullet P_C \ldots$
  $P_{OR} = P_A + P_B + P_C \ldots$ (approx)
- CSs are a key component of math
- MOEs & MOBs must be resolved

--- FT Evaluation ---

Purpose

- Obtaining the results and conclusions from the FT

- Using the FT for its intended purpose
  - Identify root causes of UE
  - Identify critical components and paths
  - Evaluate probabilistic risk

- Using the FT to impact design
  - Identify weak links
  - Evaluate impact of changes
  - Decision making

Evaluation Types

- Qualitative
  - Cut Sets

- Quantitative
  - Cut Sets
  - Probability
  - Importance Measures

Chapter 15 – Fault Tree Analysis

Methods For Finding Min CS

- Boolean reduction
- Bottom up reduction algorithms
  - MICSUP (Minimal Cut Sets Upward) algorithm
- Top down reduction algorithms
  - MOCUS (Method of Obtaining Cut Sets) algorithm
- Binary Decision Diagram (BDD)
- Min Terms method (Shannon decomposition)
- Modularization methods
- Genetic algorithms

Evaluation Trouble Makers

- Tree size
- Tree Complexity
  - Redundancy (MOEs and MOBs)
  - Large quantity of AND/OR combinations
- Exotic gates and Not logic gates
- Computer limitations
  - Speed
  - Memory size
  - Software language
- Combination of any of the above

Solutions: a) Prune FT, 2) Truncation or 3) FT Simulation

Chapter 15 – Fault Tree Analysis

CS Truncation

- Reduces number of CS's when tree is too large or complex

- Order Truncation
    - Throw away all CS's having more elements than order N_{CO}
    - Example – if N_{CO} is 3, then CS{A, B, C, D} would be dropped

- Probability Truncation
    - Throw away all CS's having probability smaller than P_{CO}
    - Example – if P_{CO} is 1.0×10^{-6}, then CS(1.0×10^{-7}) would be dropped

CS Truncation

- Reduces number of CS's when tree is too large or complex

- Order Truncation
    - Throw away all CS's having more elements than order N_{CO}
    - Example – if N_{CO} is 3, then CS{A, B, C, D} would be dropped

- Probability Truncation
    - Throw away all CS's having probability smaller than P_{CO}
    - Example – if P_{CO} is 1.0×10^{-6}, then CS(1.0×10^{-7}) would be dropped

Chapter 15 – Fault Tree Analysis

Potential CS Truncation Errors

- With Probability truncation
    - Could discard a SPF event if the probability is below the CO
- With Order Truncation
    - Could discard a significant MinCS if all the elements have a high probability
    - If algorithm used does not completely resolve CS before discarding, could miss a MOE reduction
- With either Truncation method
    - Discarded CS's are not included in the final probability
    - Must make sure the error is insignificant; accuracy is sacrificed
    - Circumvents any Common Cause analysis of AND gates

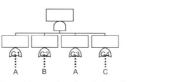

Do not truncate at gate level

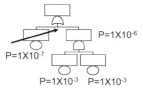

Watch for SPFs

Min CS

- A CS with the minimum number of events that can still cause the top event
- The true list of CS's contributing to the Top
- The final CS list after removing all SCS and DupCS
- Additional CS's are often generated, beyond the MinCS's
    - Super Cut Sets (SCS) – result from MOE's
    - Duplicate Cut Sets (DupCS) - result from MOE's or AND/OR combinations
- Why eliminate SCS and DupCS?
    - Laws of Boolean algebra
    - Would make the overall tree probability slightly larger (erroneous but conservative)

Chapter 15 – Fault Tree Analysis

Min CS

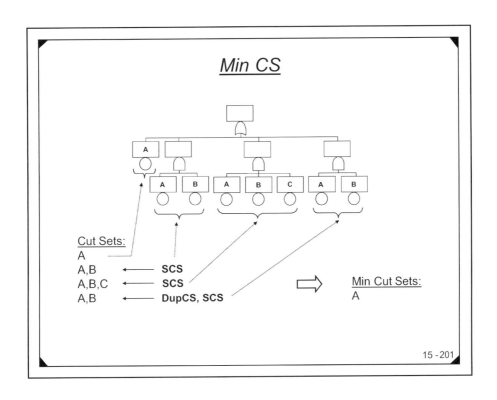

MOCUS Algorithm

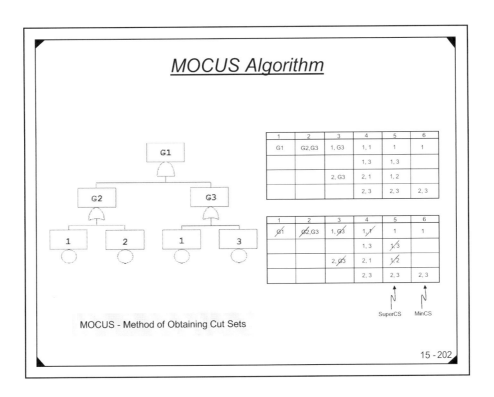

MOCUS - Method of Obtaining Cut Sets

Chapter 15 – Fault Tree Analysis

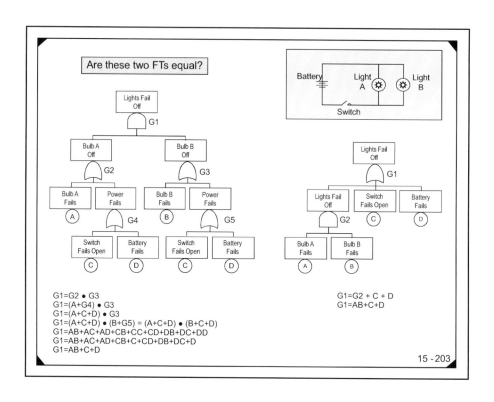

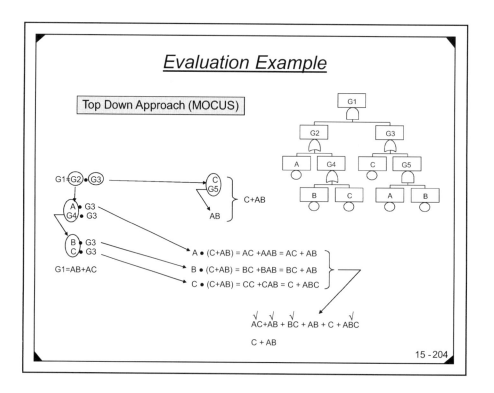

Chapter 15 – Fault Tree Analysis

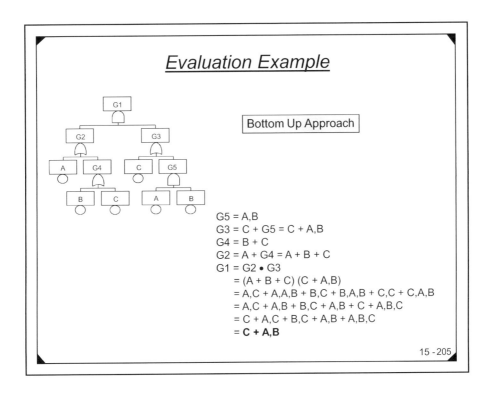

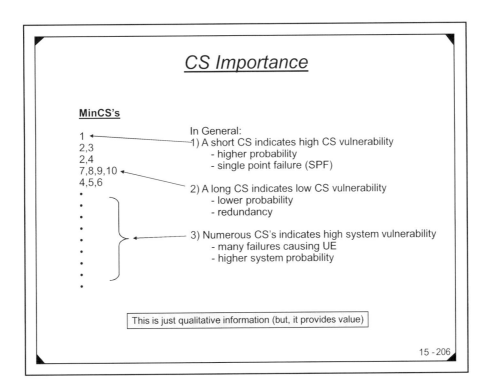

Chapter 15 – Fault Tree Analysis

Importance Measures

- Shows the relative importance of an event or CS, in comparison with the overall FT
- Help identify weak links in the system
- CS Importance Measure
    - The probability that a particular CS has caused the UE to occur
- Event Importance Measures
    - Fussel-Vesely Importance – probability of component contributing to failure of tree
    - Birnbaum Importance – if a component failure rate is changed, which component has largest impact
    - Barlow-Prochan Importance – importance of Initiator events in a repairable system
    - Barlow-Prochan Sequential Contributory Importance – importance of Enabler events in a repairable system

Cut Set Importance Measure

CS Importance

The probability that a particular CS has caused the UE to occur.

MinCSs
2
2,3
2,4 $\longrightarrow$ $I_{2,4} = \dfrac{(P_2 \cdot P_4)}{P_T}$
7,8,9,10
4,5,6

Chapter 15 – Fault Tree Analysis

Fussel-Vesely Importance Measure

$$I_{FV} = \frac{\text{Sum of unavailability of CSs containing event}}{\text{Sum of all CS unavailability}}$$

The Fussell-Vesely importance measure indicates an events contribution to the system unavailability. Increasing the availability of events with high importance values will have the most significant effect on system availability.

Item (or Event) Importance)

The probability that a particular event has caused the UE to occur.

MinCSs
2,1
2,3
2,4
7,8,9,10
4,5,6

$$I_2 = \frac{(P_2 \cdot P_1) + (P_2 \cdot P_3) + (P_2 \cdot P_4)}{P_T}$$

15 - 209

FT Quantification Overview

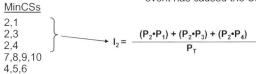

Method 1:
Event probability works up FT from bottom.

P = CS1 + CS2 + CS3 + CS4.....

Method 2:
Sum all CSs.

Both methods work good until MOEs, MOBs and large FTs are encountered.

15 - 210

Chapter 15 – Fault Tree Analysis

FT Reductions

- FT reductions are often needed

- FT reduction for tree simplification
    - Modify FT by Boolean algebra rules
    - Simplify structure and size
    - The system functional logic that derives the FT is lost

- FT reduction for MOE / MOB resolution
    - Modify Boolean equation to properly eliminate MOEs from the equation
    - Absolutely necessary for correct computations

FT Reductions

Mathematically Equivalent FTs

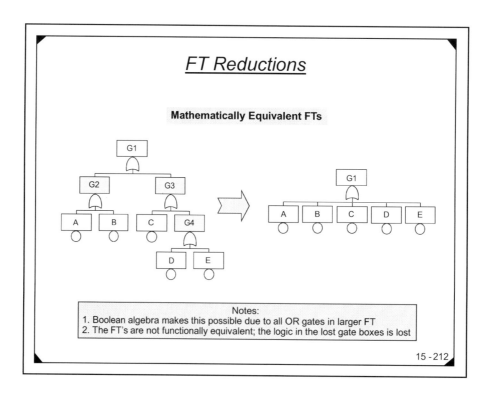

Notes:
1. Boolean algebra makes this possible due to all OR gates in larger FT
2. The FT's are not functionally equivalent; the logic in the lost gate boxes is lost

Chapter 15 – Fault Tree Analysis

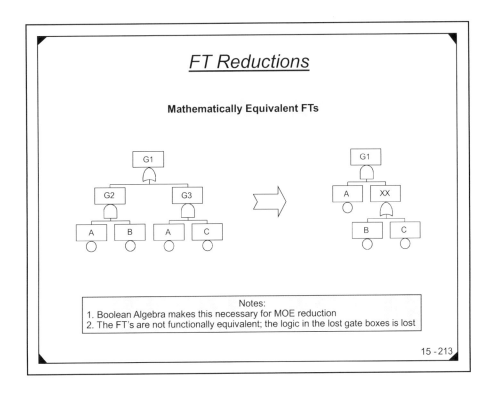

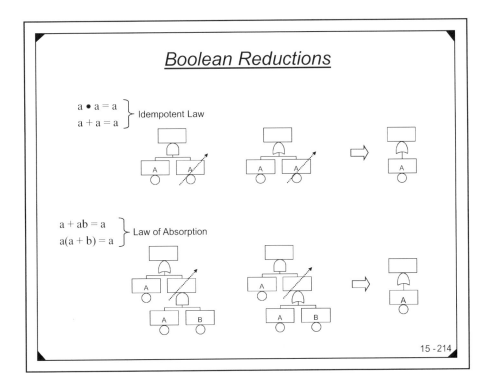

Chapter 15 – Fault Tree Analysis

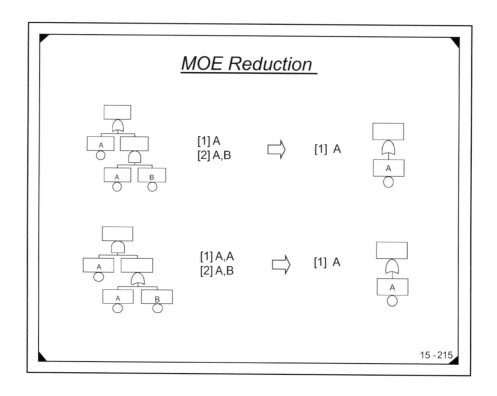
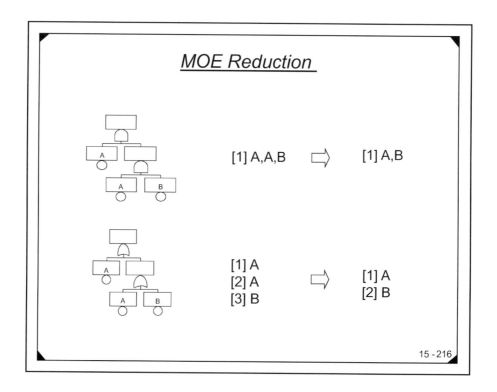

Chapter 15 – Fault Tree Analysis

CS Expansion Formula

$$P = \Sigma(\text{singles}) - \Sigma(\text{pairs}) + \Sigma(\text{triples}) - \Sigma(\text{fours}) + \Sigma(\text{fives}) - \Sigma(\text{sixes}) + \cdots$$

CS {A; B; C; D}

$$\begin{aligned}P =\ & (P_A + P_B + P_C + P_D) \\ & - (P_{AB} + P_{AC} + P_{AD} + P_{BC} + P_{BD} + P_{CD}) \\ & + (P_{ABC} + P_{ABD} + P_{ACD} + P_{BCD}) \\ & - (P_{ABCD})\end{aligned}$$

$$P = P_A + P_B + P_C + P_D - (P_{AB} + P_{AC} + P_{AD} + P_{BC} + P_{BD} + P_{CD}) + (P_{ABC} + P_{ABD} + P_{ACD} + P_{BCD}) - (P_{ABCD})$$

> Size and complexity of the formula depends on the total number of cut sets and MOE's.

Inclusion-Exclusion Approximation

CS {A; B; C; D}

$$P = \underbrace{P_A + P_B + P_C + P_D}_{\substack{\text{1st Term}\\\text{(all singles)}}} - \underbrace{(P_{AB} + P_{AC} + P_{AD} + P_{BC} + P_{BD} + P_{CD})}_{\substack{\text{2nd Term}\\\text{(all doubles)}}} + \underbrace{(P_{ABC} + P_{ABD} + P_{ACD} + P_{BCD})}_{\substack{\text{3rd Term}\\\text{(all triples)}}} - \underbrace{(P_{ABCD})}_{\text{4th Term}} \cdots$$

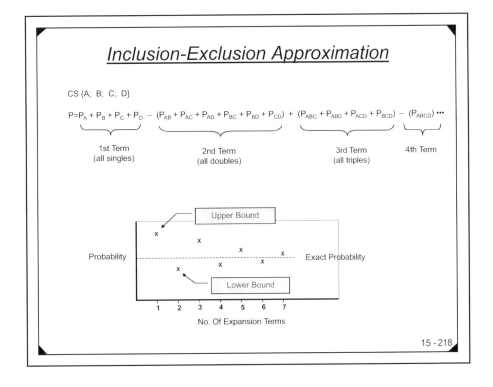

Chapter 15 – Fault Tree Analysis

Min Cut Set Upper Bound Approximation

$$P = 1 - [(1 - P_{CS1})(1 - P_{CS2})(1 - P_{CS3})\ldots(1 - P_{CSN})]$$

$CS\{A; B\}$

$$\begin{aligned}
P &= 1 - [(1 - P_A)(1 - P_B)] \\
&= 1 - [1 - P_B - P_A + P_A P_B] \\
&= 1 - 1 + P_B + P_A - P_A P_B \\
&= P_A + P_B - P_A P_B \quad \leftarrow \text{Equivalent to standard expansion}
\end{aligned}$$

FT Approximations vs. Markov

- MA
  - Small models only
  - Good numerical accuracy
  - Model is difficult to follow

- FTA
  - Defined, structured and rigorous methodology
  - Easy to learn, perform and follow
  - Provides root causes
  - Displays cause-consequence relationships
  - Sufficient accuracy when approximations are used

FTA is often criticized as not being accurate enough.

Chapter 15 – Fault Tree Analysis

Series System

Description
A system is comprised of two components A and B in series. System success requires that both must operate successfully at the same time. System failure occurs if either one or both fail.

$$P = (1 - e^{-\lambda AT}) + (1 - e^{-\lambda BT}) - (1 - e^{-\lambda AT})(1 - e^{-\lambda BT})$$
$$= 1 - e^{-(\lambda A + \lambda B)T}$$

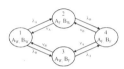

$$dP_1/dt = -(\lambda_A + \lambda_B)P_1 + \nu_A P_2 + \nu_B P_3$$
$$dP_2/dt = \lambda_A P_1 - (\lambda_A + \nu_A)P_2 + \nu_B P_4$$
$$dP_3/dt = \lambda_B P_1 - (\lambda_A + \nu_A)P_3 + \nu_A P_4$$
$$dP_4/dt = \lambda_B P_2 + \lambda_A P_3 - (\nu_A + \nu_B)P_4$$
$$P = P_2 + P_3 + P_4$$
$$P = (1 - e^{-\lambda AT}) + (1 - e^{-\lambda BT}) - (1 - e^{-\lambda AT})(1 - e^{-\lambda BT})$$
$$= 1 - e^{-(\lambda A + \lambda B)T}$$

Conclusion
Both methods produce the same results (for non-repair case).

15 - 221

Parallel System

Description
A system is comprised of two components A and B in parallel. System success requires that either one (or both) must operate successfully. System failure occurs only if both are failed at the same time.

$$P = (1 - e^{-\lambda AT})(1 - e^{-\lambda BT})$$

$$dP_1/dt = -(\lambda_A + \lambda_B)P_1 + \nu_A P_2 + \nu_B P_3$$
$$dP_2/dt = \lambda_A P_1 - (\lambda_A + \nu_A)P_2 + \nu_B P_4$$
$$dP_3/dt = \lambda_B P_1 - (\lambda_A + \nu_A)P_3 + \nu_A P_4$$
$$dP_4/dt = \lambda_B P_2 + \lambda_A P_3 - (\nu_A + \nu_B)P_4$$
$$P = P_4$$
$$P = (1 - e^{-\lambda AT})(1 - e^{-\lambda BT})$$

Conclusion
Both methods produce the same results (for non-repair case).

15 - 222

Chapter 15 – Fault Tree Analysis

Sequence Parallel System

Description
A system is comprised of two components A and B in parallel. System success requires that either one (or both) must operate successfully. System failure occurs if both fail, but only if A fails before B.

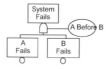

$P = (P_A \cdot P_B) / N!$
$P = (P_A \cdot P_B) / 2$
$= ((1 - e^{-\lambda AT})(1 - e^{-\lambda BT})) / 2$

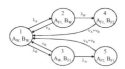

$$P = \frac{\lambda_A(1 - e^{-\lambda BT}) - \lambda_B(e^{-\lambda BT} - e^{-(\lambda A + \lambda B)T})}{\lambda_A + \lambda_B}$$

Conclusion
Each method produces a different equation, but results are comparable.

15 - 223

Sequence Parallel System

Comparison of Results for Sequence Parallel System
Where $\lambda_A = 1.0 \times 10^{-6}$ and $\lambda_B = 1.0 \times 10^{-7}$

Time (Hrs)	FTA	MA
1	5.00000E-14	5.00000E-14
10	4.99947E-12	4.99998E-12
100	4.99973E-10	4.99980E-10
1,000	4.99725E-8	4.99800E-8
10,000	4.97260E-6	4.98006E-6
100,000	4.73442E-4	4.80542E-4
1,000,000	3.00771E-2	3.45145E-2

Conclusion
Both methods produce different equations. However, for small numbers, the FTA result is a very close approximation.

15 - 224

Chapter 15 – Fault Tree Analysis

--- FT Validation ---

Purpose
- Checking the FT for errors
- Verifying the FT is correct and accurate
- Checks to convince yourself that the tree is correct

Why
- Very easy to introduce errors into the FT
- FT misuse and abuse is very easy
- Helps to ensure the results are correct
- Helps to reassure the customer
- Helps to reassure management

Validation is probably one of most ignored steps in the FTA process

How Errors Are Introduced

- Analyst does not understand system
- Analyst does not fully understand FTA
- FTs can become very complex
  - Modeling a complex system design
  - FT understanding can decrease as FT size increases
- Communication errors between several FT analysts
- Errors in tree structure logic sometime occur (wrong gate selected)
- A MOE component is given the wrong name (it's not really a MOE)
- Computer evaluation codes are erroneous
- Computer evaluation codes are used incorrectly
- Incorrect (or out of date) system data is used
  - Failure rates, drawings, design data

Chapter 15 – Fault Tree Analysis

Methods For Validating The FT

- CS reality check
- Probability reasonableness test
- Success tree inversion
- Gate check
- UE check
- Check against definitions and boundaries
- Review of failure data
- Metrics check
- Peer review
- MOE check
- Intuition check
- Logic Loop check

--- FTA Audit ---

- Purpose – To verify and validate a contractual FTA product

- To evaluate an existing FT for:
    - Correctness
    - Completeness
    - Thoroughness

- To determine if the results from a FTA are valid
    - Determine if the FTA contains defects
    - Avoid making decisions on incorrect analysis results

Chapter 15 – Fault Tree Analysis

Audit vs. Validation

- A FTA audit is similar to FT validation, but not the same
- FT Audit
  - Typically performed by independent reviewer after FTA is complete
  - Auditor may not have all detailed design information or knowledge
- FT Validation
  - Typically performed by the product developer
  - Analyst has detailed design information
- Validation items that can be used for audit
  - CS reality check
  - Probability reasonableness test
  - Gate check
  - Review of failure data
  - MOE check

Audit Guidelines

- Need a basic understanding of system design (optimally)
- Evaluate FT for each potential defect category
- Question everything
  - Check with SME if possible
- If something looks funny, it probably is
- Documents audit data and results

A FT auditor:
- Must understand FT construction thoroughly
- Must be a highly experienced FT analyst

Chapter 15 – Fault Tree Analysis

Defect Categories

- Math
- Fault logic
- Failure data
- Evaluation methods
- Completeness
  - Anything omitted
- Analysis ground rules
  - Are rules established and followed?
  - Rules on SW, HSI, CCF, exposure time, depth of analysis
- Diagramming
  - Symbol use
  - Aesthetics

FTA Error/Defect Levels

Error	Consequence
High	Erroneous top probability
Medium	Insufficient Info; not sure if results are incorrect
Low	Poor FT diagram, however, results are likely correct

Chapter 15 – Fault Tree Analysis

High Consequence Errors

- Gate logic error
  - AND vs. OR, logic does not correctly model system design, etc.
  - House event into an OR gate

- Omitting necessary design detail
  - Subsystems
  - Human error, SW, HSI interface design

- Cut set errors
  - Incorrect, missing, contradicting

- Mathematical errors
  - MOE resolution error, calculation error, normalizing error

- Input data errors
  - Incorrect failure rate or time

- Common FT pitfall type errors
  - Extrapolation, dependency, truncation, mutual exclusion, latency, CCF

Medium Consequence Errors

- Inadequate information errors
  - Missing text description (e.g., "Resistor fails" – open, short, tolerance?)
  - Vague text description (e.g., "Spring way too strong")

- Missing information
  - Blank text boxes
  - Missing text boxes

- Jumping ahead in system fault path
  - Skipping fault logic steps

- Manipulations used to obtain favorable probability results

- Failure data
  - No reference sources
  - Failure rates are questionable (reasonable?)

Chapter 15 – Fault Tree Analysis

Low Consequence Errors

- Violation of common FT rules
    - Gate to gate
    - Multiple outputs (double connects)
    - No text in boxes
    - Inputs/outputs on side of box (vice top/bottom)
    - Incorrect symbol usage
- FT Sloppiness
    - Messy diagram
    - Unreadable text (hand drawn)
    - Too small to read

Audit Checklist

- Do CSs make sense and do they cause UE
- Are all of the CSs minimal
- Are any CSs mutually exclusive
- Is the Probability reasonable (based on data and experience)
- Do Gates appear correct
- Is failure data reasonable
- Are MOEs and MOBs correct
- Does FT diagram follow basic rules
- Do all nodes have text boxes with words
- Does wording in text boxes make sense
- Does the overall fault logic seem reasonable
- Is the math correct
- Has latency been considered
- Has common cause been considered
- Has human error been considered
- Are the component exposure times correct

Chapter 15 – Fault Tree Analysis

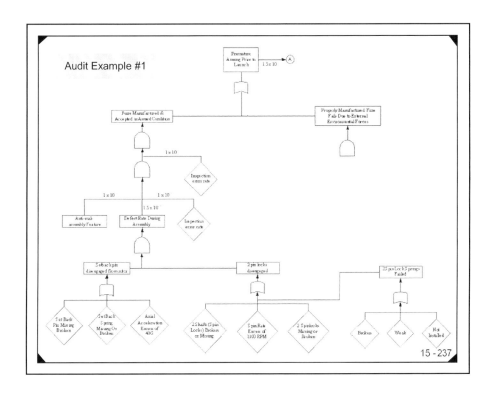

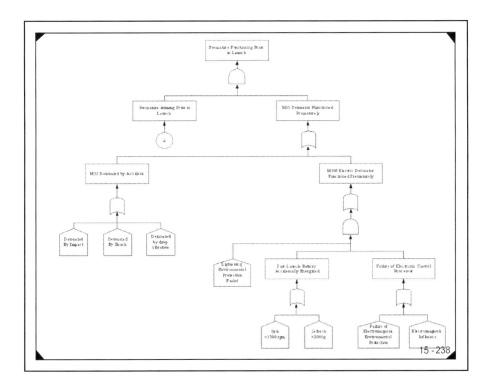

Chapter 15 – Fault Tree Analysis

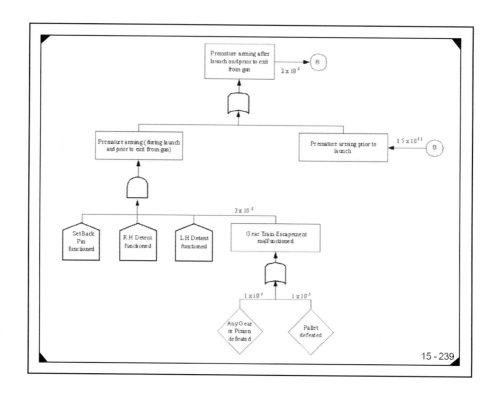

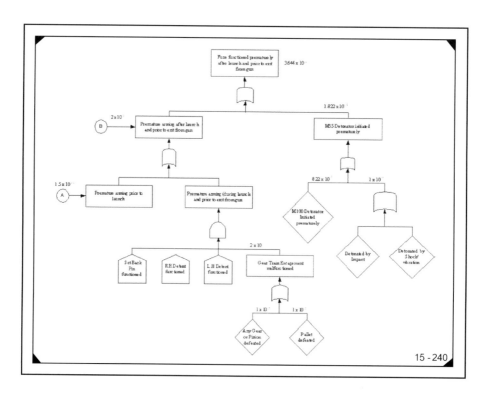

Chapter 15 – Fault Tree Analysis

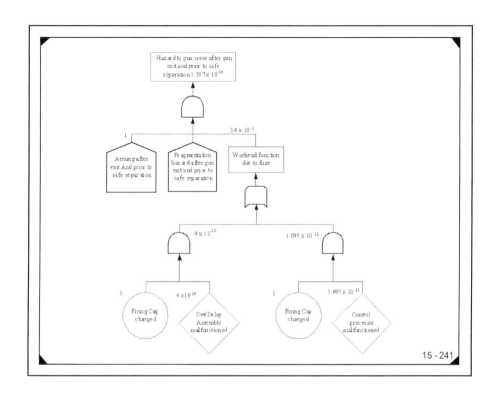

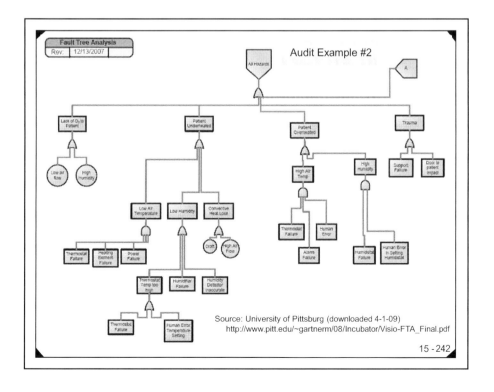

Chapter 15 – Fault Tree Analysis

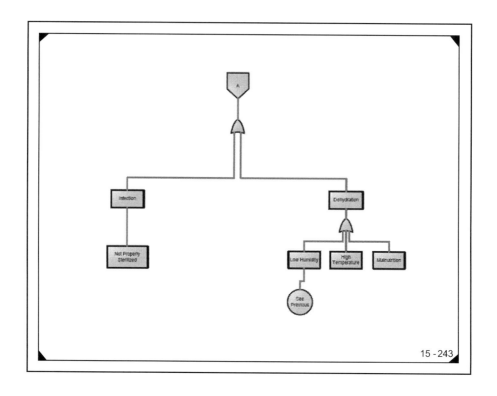

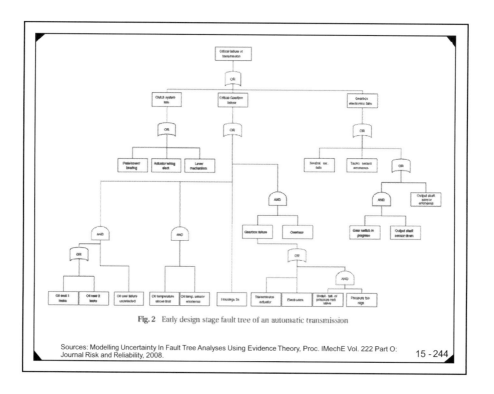

Fig. 2 Early design stage fault tree of an automatic transmission

Sources: Modelling Uncertainty In Fault Tree Analyses Using Evidence Theory, Proc. IMechE Vol. 222 Part O: Journal Risk and Reliability, 2008.

Chapter 15 – Fault Tree Analysis

Audit Guidance

- If you have received a FTA from a contractor or supplier, it's important to obtain an independent audit of the analysis
  - Ensure the probabilities you are basing decisions on are correct
  - Require a written audit report

- Check for three defect categories
  - High, Medium and Low

- Any defects in the High category mean the FT is incorrect

--- Misc FTA Aspects ---

Latency

- Latency refers to a latent component failure, which is a component could be failed for some time without knowledge.

- A Latent component is a component that is not checked for operability before the start of a mission. Thus, it could already be failed at the start of the mission.

- This effectively increases the component exposure time. The latent time period is the time between checks (ie, Maintenance), which can often be significantly greater than the mission time. This large exposure time can make a large impact on the probability.

Chapter 15 – Fault Tree Analysis

Latency

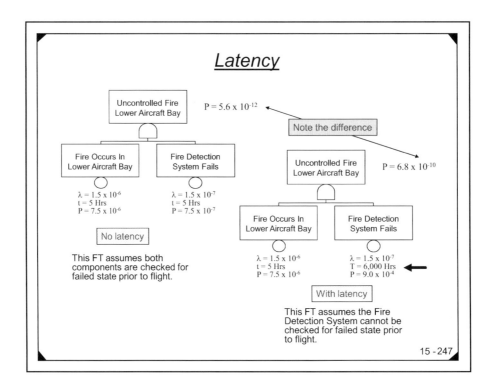

CCF

- A Common Cause Failure (CCF) is a single point failure (SPF) that negates independent redundant designs
- Typical CCF sources
  - Common weakness in design redundancy
    - Example – close proximity of hydraulic lines
  - The use of identical components in multiple subsystems
  - Common software design
  - Common manufacturing errors
  - Common requirements errors
  - Common production process errors
  - Common maintenance errors
  - Common installation errors
  - Common environmental factor vulnerabilities

Chapter 15 – Fault Tree Analysis

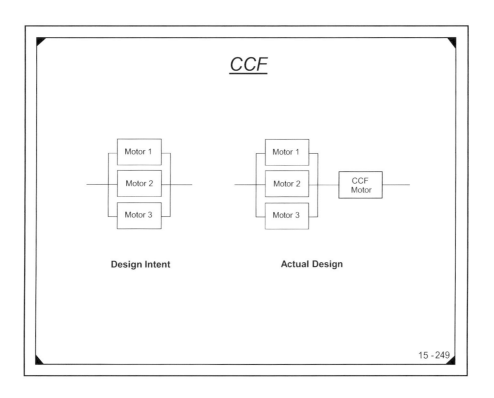

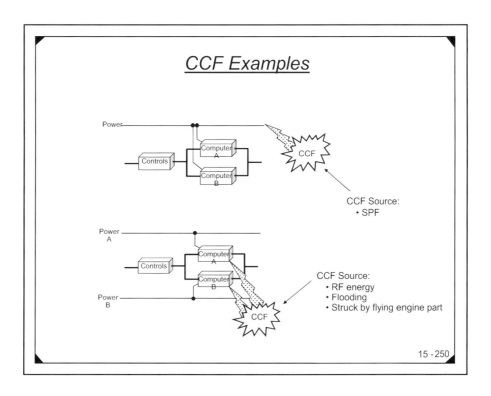

Chapter 15 – Fault Tree Analysis

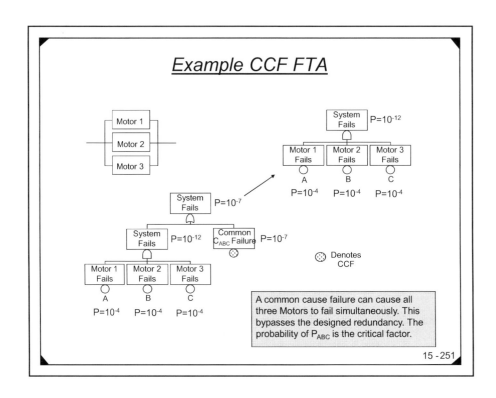

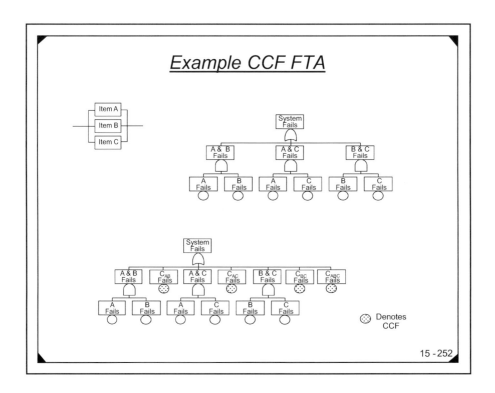

Chapter 15 – Fault Tree Analysis

Interlocks

- An interlock is usually designed into a system for one of two purposes:
  - To help prevent the inadvertent operation of a critical function
  - To stop the operation of a hazardous operation or function before a mishap occurs
- A safety interlock is a single device that is part of a larger system function; only necessary for safety, not function (enable)
- Its purpose is to prevent the overall system function from being performed until a specified set of safety parameters are satisfied.
- An interlock can be implemented in either hardware or software

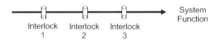

Interlocks

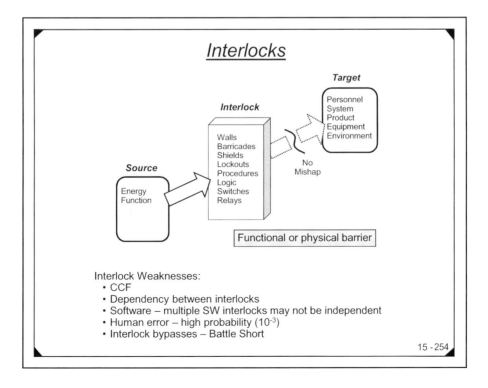

Interlock Weaknesses:
- CCF
- Dependency between interlocks
- Software – multiple SW interlocks may not be independent
- Human error – high probability (10^{-3})
- Interlock bypasses – Battle Short

Chapter 15 – Fault Tree Analysis

Interlock Example

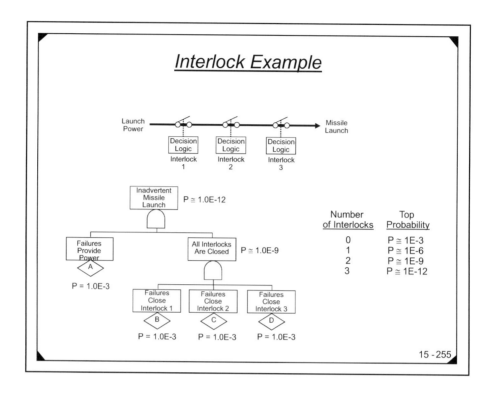

Number of Interlocks	Top Probability
0	$P \cong 1E\text{-}3$
1	$P \cong 1E\text{-}6$
2	$P \cong 1E\text{-}9$
3	$P \cong 1E\text{-}12$

15 - 255

Dependency

- An Independent event is an event that <u>is not</u> influenced or caused by another event
- A Dependent event is an event that <u>is</u> influenced or caused by another event
- Dependencies complicate FT math considerably
  - Conditional probability
  - Requires Markov analysis for accuracy
  - However, FT approximations are quite accurate
- Sometimes dependencies are difficult to identify
  - A Secondary failure may or may not be the cause of a dependent failure
  - If A causes B, then in this case Prob(B/A) should be more likely than independent Prob(B)
  - Secondary RF energy may cause a transistor to fail, but they are "typically" considered independent (the approximation is accurate enough)

15 - 256

Chapter 15 – Fault Tree Analysis

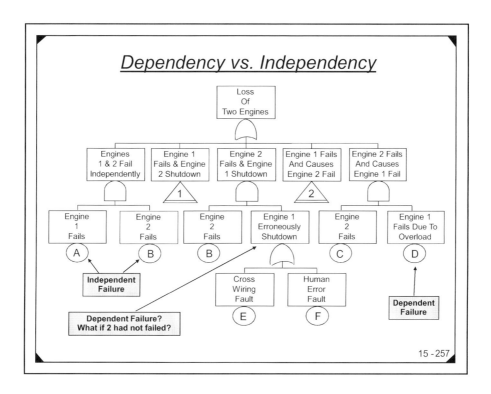

RTCA DO-178B

- Software level is based upon the contribution of software to potential failure conditions as determined by the system safety assessment process (SSAP).
- The software level implies that the level of effort required to show compliance with certification requirements varies with the failure condition category.

Level	Definition
A	Software whose anomalous behavior, as shown by the SSAP, would cause or contribute to a failure of system function resulting in a catastrophic failure condition for the aircraft
B	Software whose anomalous behavior, as shown by the SSAP, would cause or contribute to a failure of system function resulting in a hazardous/severe-major failure condition of the aircraft
C	Software whose anomalous behavior, as shown by the SSAP, would cause or contribute to a failure of system function resulting in a major failure condition for the aircraft
D	Software whose anomalous behavior, as shown by the SSAP, would cause or contribute to a failure of system function resulting in a minor failure condition for the aircraft
E	Software whose anomalous behavior, as shown by the SSAP, would cause or contribute to a failure of function with no effect on aircraft operational capability or pilot workload

Chapter 15 – Fault Tree Analysis

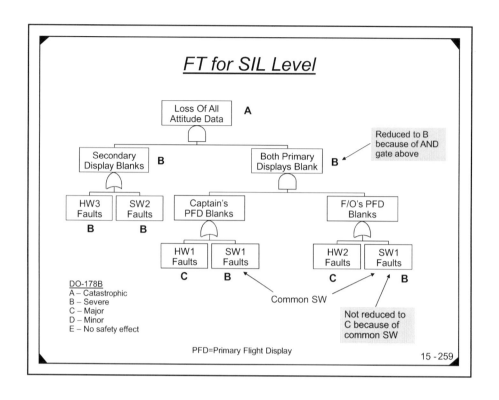

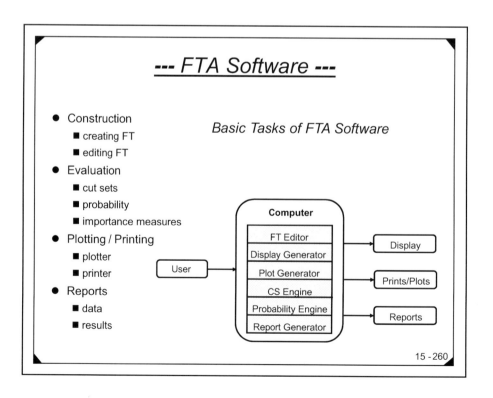

Chapter 15 – Fault Tree Analysis

Properties of FT Software

	Required	Desired
• Construction (create, edit, copy, past)	X	
• Generate Min CS	X	
• Generate probabilities (top, gates)	X	
• Employ CS cutoff methods (order, probability)		X
• Numerical accuracy	X	
• Generate reports	X	
• Detect tree logic loops		X
• Data export capability (tree structure, failure data)		X
• Graphic export capability (BMP, JPG)		X
• Tree Pagination for prints	X	
• User select print size		X
• Program fit on a floppy disk		X
• Notes on FT print/plot		X
• Find feature		X
• Undo feature		X

Properties (continued)

	Required	Desired
• Employ MOEs / MOBs	X	
• Correctly resolve MOEs / MOBs	X	
• Global data change (failure rates, exposure time)		X
• Print/plot selected pages or all pages	X	
• Unrestricted FT tree size	X	
• Automatic naming of gates and events when created	X	
• Capability to resolve large complex FTs	X	
• User friendly (intuitive commands)	X	
• Print/plot results visually aesthetic		X
• Verification of mathematical methods and accuracy	X	
• Open data file structure	X	

Chapter 15 – Fault Tree Analysis

Commercial FT Software

CAFTA	CAFTA sales 650-384-5693 (part of EPRI)
Fault Tree+	www.Isograph.com
Item	www.itemsoft.com
Relex	www.relexsoftware.com
FaulTrEase	http://www.chempute.com/faultrea.htm
Risk Spectrum FT	http://www.relcon.com
Shade Tree	http://www.qrainc.com
Tree-Master	http://www.mgtsciences.com
FTA Pro	http://www.dyadem.com
OpenFTA (free source)	http://www.openfta.com
Logan	http://www.arevarmc.com/logan-faulttreeanalysis.php
RAM Commander	http://www.aldservice.com
Saphire (free?)	https://saphire.inl.gov/faq.cfm

Purchase Considerations

- Many commercial FT codes are available -- some good and some ?
- Know FT tool limits and capabilities
    - Tree size, Cut set size, Print size, numerical accuracy
    - Easy to use (without tech support)
    - Single phase, multi-phase
- Understand algorithms
    - Some codes have errors in approximations and cutoff methods
    - Gate probability calculations must resolve MOEs correctly
- Test the tool; don't assume answers are always correct
- Consider
    - Price
    - Lease vs. Ownership
    - User flexibility – networks, stations, individuals
    - Maintenance
    - Lease / Buy
    - User friendliness
    - Training

Chapter 15 – Fault Tree Analysis

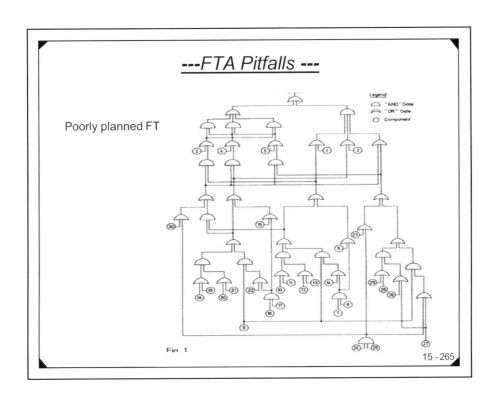

---FTA Pitfalls ---

Poorly planned FT

Fig. 1

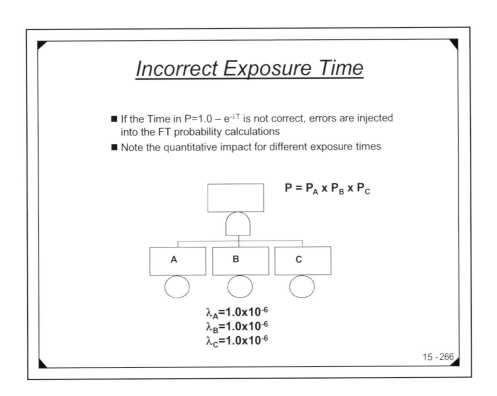

Incorrect Exposure Time

- If the Time in $P = 1.0 - e^{-\lambda T}$ is not correct, errors are injected into the FT probability calculations
- Note the quantitative impact for different exposure times

$$P = P_A \times P_B \times P_C$$

$\lambda_A = 1.0 \times 10^{-6}$
$\lambda_B = 1.0 \times 10^{-6}$
$\lambda_C = 1.0 \times 10^{-6}$

Chapter 15 – Fault Tree Analysis

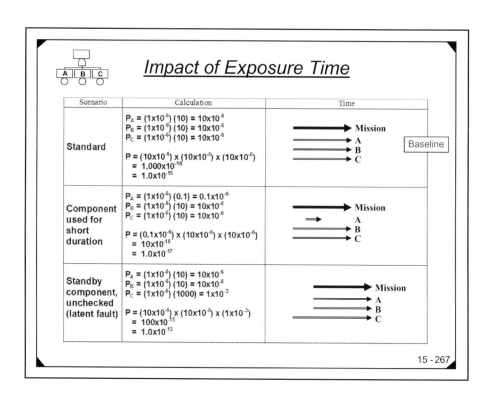

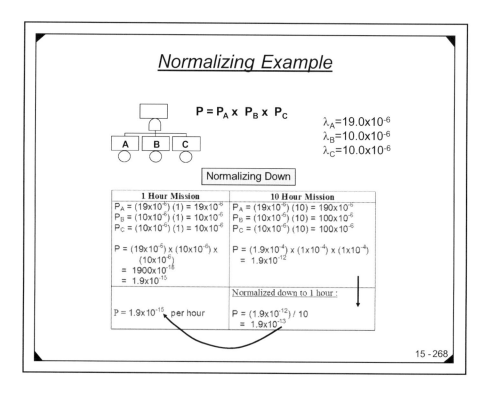

Chapter 15 – Fault Tree Analysis

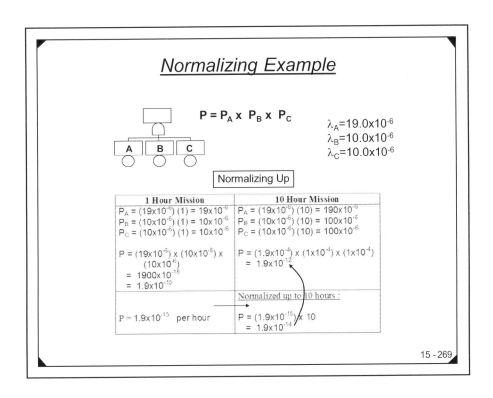

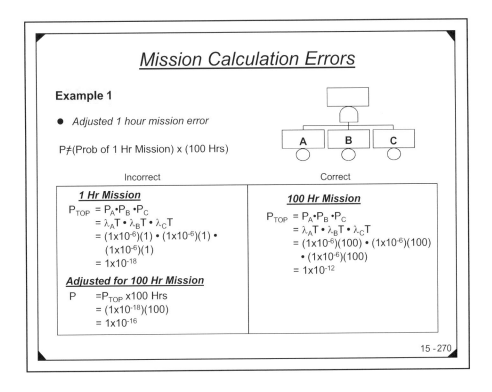

Chapter 15 – Fault Tree Analysis

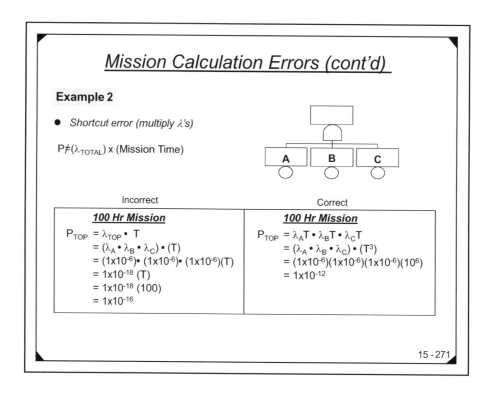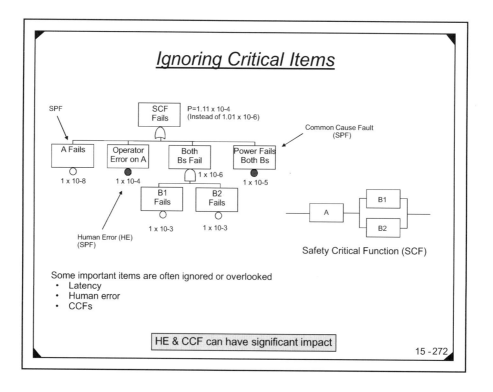

Jumping Ahead

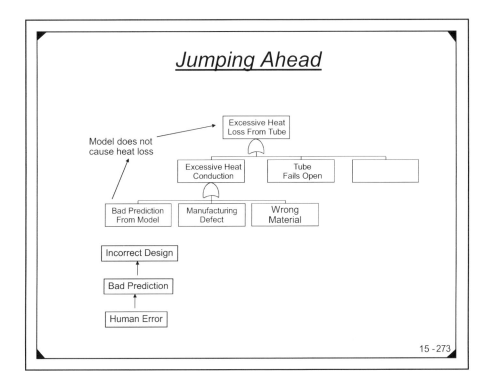

15 - 273

Summary

- Types of FT evaluation
  - Qualitative
    - CSs show design weaknesses
  - Quantitative
    - CSs show design weaknesses
    - Probability calculations provide numerical estimates
      * CSs
      * Gates (FT branches)
      * UE (total FT)
      * Importance Measures
      * Possible design errors (CCFs, SPFs, Interlocks, etc.)
- CSs are a key product of the FT
- When MOEs and MOBs exist in FT, Boolean reduction must be done properly to avoid incorrect results
- AND gates reduce probability
- OR gates increase probability (and computation complexity)

15 - 274

Chapter 15 – Fault Tree Analysis

System Safety and Reliability Analysis
Course Notes

Chapter 16
Software Safety

Clifton A. Ericson II
Design Safety Solutions LLC
cliftonericson@verizon.net
540-786-3777

© C. A. Ericson II 2014

Introduction

- Software (SW) related mishaps are real
  - Many mishaps have been attributed to SW

- The safety threat presented by SW is growing
  - Increased usage of SW in system design
  - Increased system control given to SW
  - As system complexity increases, so does SW size and complexity

- Software Safety (SwS) is the process/discipline for developing safe software
  - SwS is part of system safety – with differences

- SwS is more than looking for bugs, it is looking for SW related hazards

Chapter 16 – Software Safety

Concerns

- Software use is becoming more prevalent in all aspects of life
    - Software usage in systems is growing at a high rate (aircraft, cars, appliances)

- Control
    - Software is being given ultimate decision authority
    - System operation is becoming dependent on software

- Complexity
    - Too large and complex to fully understand
    - Too large and complex for a simple tool

- Risk
    - More and more systems using software
    - Statistically increases accident probability without special safety measures
    - Mishap potential not always well understood

Problems

- Software characteristics are unique and problematic
    - Software complexity is unique
    - Software development methods are less than adequate
    - COTS software is becoming prevalent
    - COTS software typically has unknown safety pedigree
    - There are too many paths in software for complete testing
    - Software testing only shows the existence of problems, it does not guarantee the absence of problems
    - Software does exactly as programmed, good or bad

- Software is abstract and difficult to fully understand

Chapter 16 – Software Safety

Software's Capabilities

- Software can cause improper or erroneous hardware operation.
- Software can overstress hardware
- Software can change hardware dependent timing
- Software can give erroneous data to hardware/ operator
- Software can erroneously interpret data from hardware/operator
- Software can perform unique unintended functions
- Software can provide inadvertent output signals
- Software can provide erroneous output signals

Example SW Mishaps

- B-1 Bomber Maintenance Incident (1978)
- Therac-25 Radiation Therapy Machine (1985-1987)
- Patriot Missile (1991)
- Lufthansa Flight 2904 (1993)
- Ariane 5 Rocket (1996)
- Predator UA Mishap (2006)
- V-22 Osprey tilt-rotor helicopter (2001)
- Toyota cars experience unintended acceleration (2010)

Chapter 16 – Software Safety

SwS Definition

- SwS is an engineering methodology employed to intentionally ensure that software is safe for use in a system context.
- SwS is subset of the system safety process.
- Software safety proactively designs safety into the system and establishes a safety case verifying the design safety tasks performed.
- Software safety is based on systems thinking, which concentrates on the design and application of the whole system as distinct from the individual parts.

Definitions

- Defect
    - A condition or characteristic in any hardware or software which is not in compliance with the specified configuration, or design requirements. A defect may exist without leading to a failure or a hazard.
- Software Error
    - The difference between a computed, observed, or measured value or condition and the true, specified or theoretically correct value or condition.
- Failure
    - A failure is the inability of an item to perform its required functions within specified performance requirements.
    - HW has physical failure modes
    - SW has functional failure modes

Chapter 16 – Software Safety

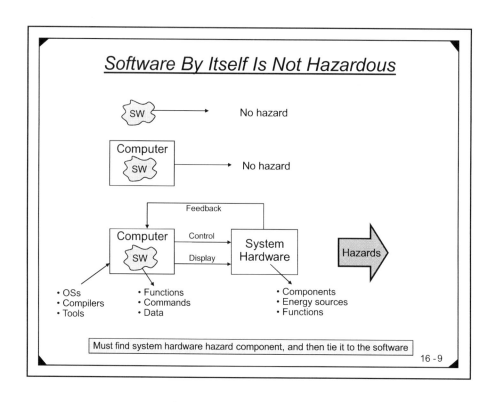

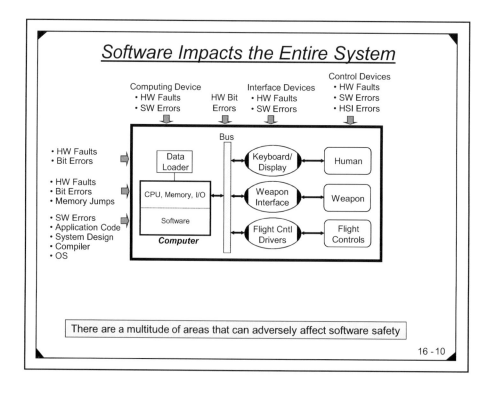

Chapter 16 – Software Safety

SwS Requires A Special Process

- SW must be treated differently in the area of safety
  - Complexity
  - Unique characteristics
  - Subtle hazard causal factors
  - Difficult to totally understand
  - Unintended functions
  - Lack of failure rates
  - Requirement gaps

- Other methods such as reliability, quality, etc. do not adequately and thoroughly affect the safety property

- A systems approach called *software safety* is required

Software Safety - History

- Software safety has been a major concern since circa 1972
  - government / manufacturers / contractors / public
- Many techniques have been proposed (few have stuck)
- Some accidents have been software caused
- Everyone is still searching for the "silver bullet" (easy out!)
- Still basically doing what was proposed in 1980, with some enhancements:
  - JSSS Handbook
  - STANAG-4404 / AOP-52
  - DO-178C
  - MIL-STD-882E

Chapter 16 – Software Safety

Software May Have Ultimate Control

"You've got to hand it to those computers!"

The Basic Problem

- Software is different than hardware
- Software has some unique characteristics
- Software use is increasing
- SW is being given absolute control over safety critical systems
- Software increases system complexity
- As software size increases it is less well understood
- Often SwS is ignored because of the difficulty it presents
- To date, SwS methods have been lacking

Chapter 16 – Software Safety

Unique SW Features (1)

- Wear-out
  - SW does not wear-out over time from ageing or use, as does hardware
- Degradation
  - SW can experience various forms of degradation as the result of poor coding (e.g., memory deterioration)
- Failure Modes
  - SW has abstract functional failure modes whereas HW has discrete physical failure modes
- Failure Rates
  - HW has known failure modes with statistically predictable failure rates, whereas SW does not
- Reliability Prediction
  - HW reliability can be predicted and measured, whereas SW reliability is much more difficult to predict and measure

Unique SW Features (2)

- Parts Reliability
  - HW parts can be purchased based on their reliability, and when necessary high reliability parts can be obtained, whereas software has no similar concept
- Modification
  - SW can be modified easier than hardware; invites first place to modify, but can be more costly
- Abstraction Level
  - SW is more abstract (intangible) compared to HW which is more concrete (tangible)
- Paths
  - SW contains more distinct internal paths than HW, making 100% testing of software difficult or impossible
- COTS
  - Cots HW tends to be more effective than COTS SW; COTS SW has many unknowns

Chapter 16 – Software Safety

Unique SW Features (3)

- Repair
  - Repair of HW restores the hardware to a new original condition with a fresh new failure rate, whereas SW repair creates a new software baseline with potentially more errors injected
- Complexity
  - SW has a greater complexity level than HW, making understanding harder
- States
  - SW can involve or generate more system states than HW, making understanding harder
- Self-Modification
  - SW can be dynamically modified during operation by HW bit faults, thereby causing the system to enter unknown states
- Hazards
  - HW alone can create hazards, whereas SW by itself cannot; SW must be combined with HW in order to be hazardous

Unique SW Features (4)

- Errors
  - SW can contain errors that may not be apparent or encountered for many years, whereas HW failures generally become apparent immediately
- Not all SW errors (defects) cause safety problems
- Reliable SW does not necessarily make or guarantee safe SW
- SW requirements do not necessarily make or guarantee safe SW
- SW always works as designed, unless it encounters unplanned factors
- SW may contain unintended functions (which are also unknown)
- SW language syntax can lead to many safety problems (e.g., ptrs)

Chapter 16 – Software Safety

Software Myths

- Computers reduce risk over mechanical systems
- Reusing software saves money and improves safety
- Increasing software reliability will improve safety
- Removing all software bugs will provide 100% safety
- Formal proofs of requirements will provide 100% safety
- A good SW design development process will resolve all problems

SW Complexity Issues (1)

- Software is written in a programming language
  - abstract
  - logic statements
  - many possible paths of operation
- Software is designed to perform diverse functions (via hardware)
  - contributes to complexity
- Software has no failure modes
  - errors
  - HW induced faults
- Software is only hazardous when:
  - controlling hardware
  - performing functions via hardware

Chapter 16 – Software Safety

SW Complexity Issues (2)

- SW risk assessment
  - Hazard mitigation is based on hazard risk (R= P x S)
  - The risk of HW caused hazards can be established via HW failure rates
  - However, the risk of SW related hazards cannot be established because SW probabilities do not exist.

- SW hazard causal factors
  - SW size, complexity and abstractness make code difficult to analyze
  - Thus, hazard causal factors can be difficult to identify
  - More stringent analysis and testing is required
  - Level of effort is often based on SW criticality (integrity level)

SW Characterization Models

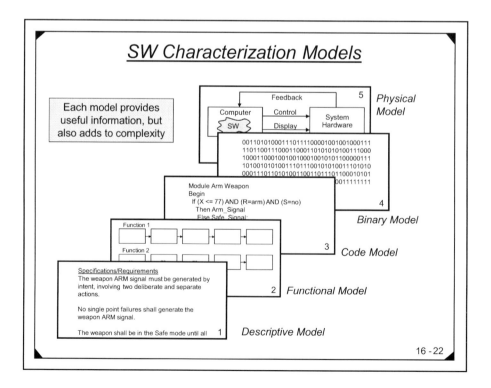

Each model provides useful information, but also adds to complexity

5 — Physical Model
4 — Binary Model
3 — Code Model
2 — Functional Model
1 — Descriptive Model

Chapter 16 – Software Safety

Design Complexity Factors

- Partitioning
- Redundancy
- Concurrency
- Dissimilar code
- Language capabilities
- Floating point vs. integer math
- Fault tolerance
- Safety Kernels
- Operating systems

SW Development Issues

- Who is responsible for safe software
  - Design
  - Safety
  - System specifications
- Definitions and terms confusion
  - Failure / Error / Hazard
  - Safe / Trustworthy / Correct
- Development tools
  - Design
  - Engineering analysis
  - Computer science/mathematical

Chapter 16 – Software Safety

SwS Dilemma

- Safety is typically achieved by hazard control:
  - Identifying hazards
  - Eliminating hazards
  - Mitigating hazard risk through design measures
    - Design-in hazard defenses
    - Mitigate until risk is acceptable
  - Verifying mitigation success

- However, this approach does not work for SW hazards
  - It's not possible to identify all SW hazards
  - Hazard SW causes become blurred (i.e., unintended functions)
  - Unforeseen flawed SW design results in unforeseen hazards

- A two-pronged approach is needed: hazard control AND software development rigor

SwS Problem #1

- Hazards can be postulated and proven to exist (or not) for:
  - Hardware
  - People
  - Environment

- Specific causal factors (CFs) are identified
  - Failure modes, interfaces errors, human errors, dependencies, etc.

- However, SW hazards typically cannot be proven to exist (by analysis)
  - SW does not have real failure modes
  - Specific causal factors cannot be identified
  - SW code is too large, complex and abstract
  - Many critical interactions and dependencies are not foreseen

Hazards can be foreseen, but SW CFs cannot be firmly identified

Chapter 16 – Software Safety

SwS Problem #2

- Hazard risk can be predicted for:
    - Hardware
    - People
    - Environment
- Risk is product of hazard Likelihood x hazard Severity
    - HW failure modes and human errors have probabilities
- SW hazards do not have a probability component
    - SW does not have real failure modes or failure rates
    - SW hazard risk cannot be determined
    - SW module safety risk cannot be determined
- Therefore, SW hazard risk cannot be measured
    - Mitigation and acceptability cannot be measured

Hazard risk can be predicted, except for SW hazard risk

16 - 27

SwS Process

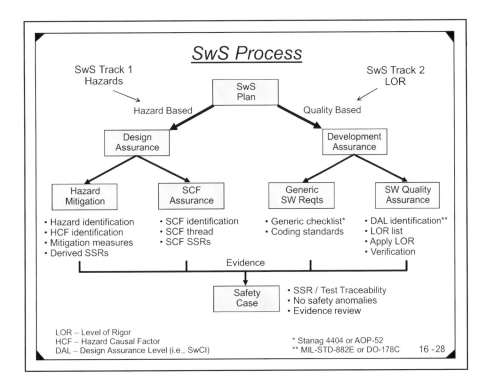

LOR – Level of Rigor
HCF – Hazard Causal Factor
DAL – Design Assurance Level (i.e., SwCI)

* Stanag 4404 or AOP-52
** MIL-STD-882E or DO-178C

16 - 28

Chapter 16 – Software Safety

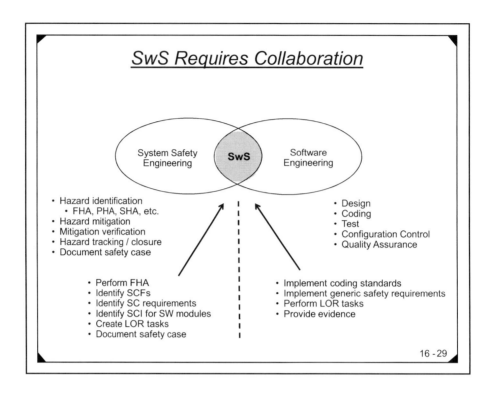

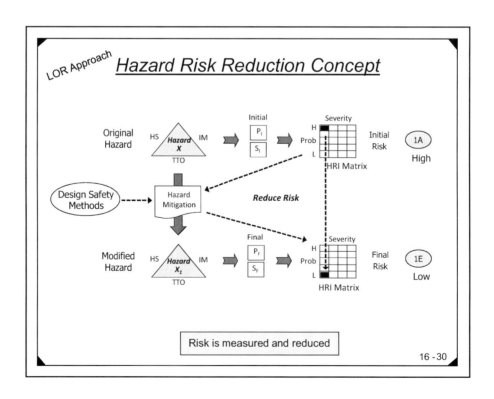

Chapter 16 – Software Safety

Risk Cannot Be Calculated For SW Hazards

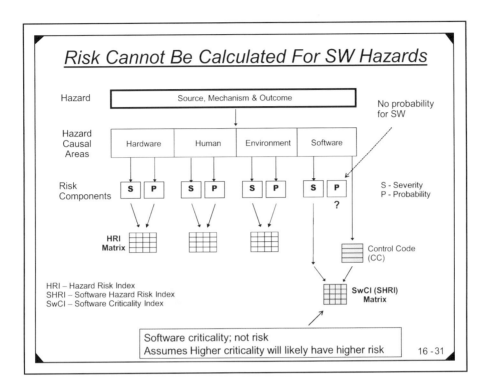

HRI – Hazard Risk Index
SHRI – Software Hazard Risk Index
SwCI – Software Criticality Index

Software criticality; not risk
Assumes Higher criticality will likely have higher risk

16 - 31

SW Module Risk

- SW risk cannot be determined
  - SW hazards
  - SW modules
- Since SW risk cannot be evaluated, a new approach is needed
  - SW criticality
  - The more critical the SW, the higher risk it could present
  - This inherent risk can be reduced via development rigor
    - The more critical the SW, the more development rigor tasks must be performed to assure no safety problems exist
    - The more rigor applied, the higher the confidence is that the SW presents low risk
- This approach involves
  - SwCI
  - SW risk criticality matrix
  - LOR tasks table

16 - 32

Chapter 16 – Software Safety

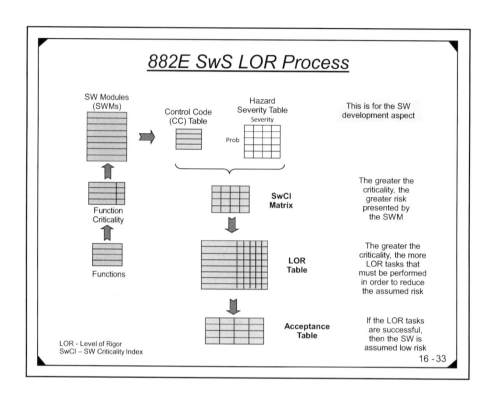

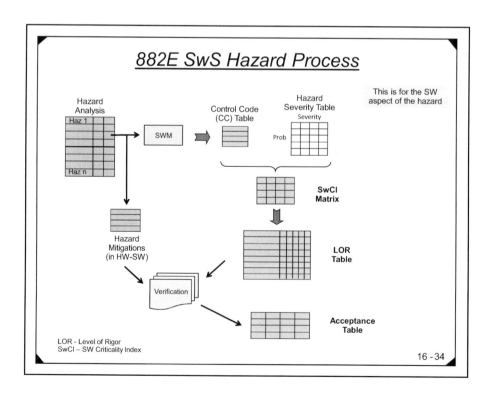

Chapter 16 – Software Safety

SwCI / LOR Concept

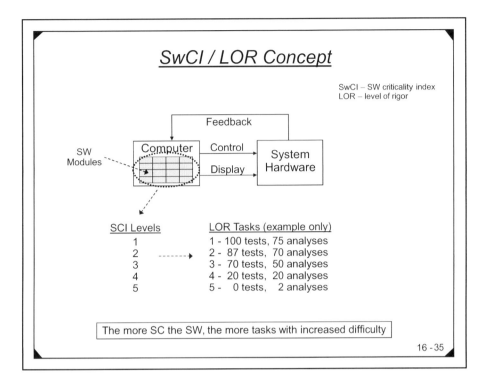

SwCI – SW criticality index
LOR – level of rigor

SCI Levels
1
2
3
4
5

LOR Tasks (example only)
1 - 100 tests, 75 analyses
2 - 87 tests, 70 analyses
3 - 70 tests, 50 analyses
4 - 20 tests, 20 analyses
5 -   0 tests, 2 analyses

The more SC the SW, the more tasks with increased difficulty

Software Criticality Index (SwCI)

- The SwCI rates the relative safety risk a SW module appears to present, based on overall functional design
- Once established, the SwCI of a SW module cannot ever change (unless the functional design concept changes)
- The SW module is developed to the LOR attached to the SwCI
- The SW risk is presumed safe after LOR tasks are performed

- What viewpoint is the SW SwCI rated from?
    - SW module standalone
    - SW module involved in SCF thread
    - SW module involved in a hazard
    - SW module involved in SC requirement

Chapter 16 – Software Safety

SwCI

- SwCI establishes the level of safety integrity for a SW module

- Note that this is a SW module criticality risk not a hazard risk

- A low SwCI value does not mean the module is unsafe, it indicates that the module is very safety related and should be treated with greater respect (test and analysis)

16 - 37

HRI Concept for Hazard

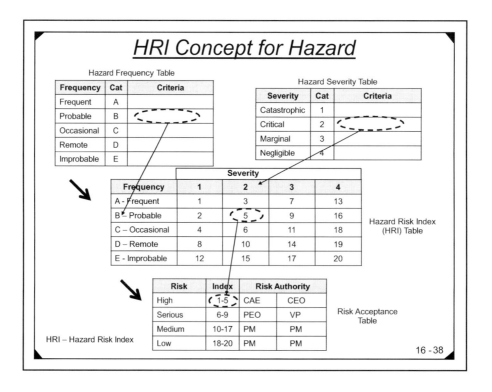

HRI – Hazard Risk Index

16 - 38

Chapter 16 – Software Safety

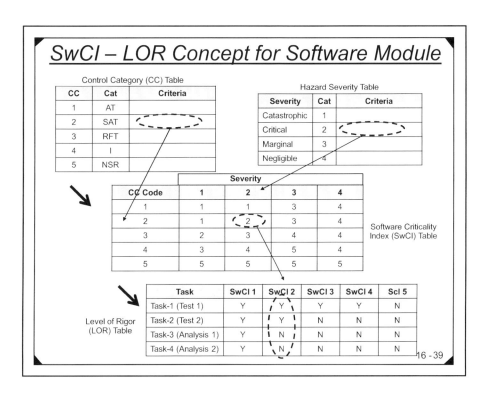

SwCI – LOR Concept for Software Module

Control Category (CC) Table

CC	Cat	Criteria
1	AT	
2	SAT	
3	RFT	
4	I	
5	NSR	

Hazard Severity Table

Severity	Cat	Criteria
Catastrophic	1	
Critical	2	
Marginal	3	
Negligible	4	

Software Criticality Index (SwCI) Table

CC Code	Severity 1	2	3	4
1	1	1	3	4
2	1	2	3	4
3	2	3	4	4
4	3	4	5	4
5	5	5	5	5

Level of Rigor (LOR) Table

Task	SwCI 1	SwCI 2	SwCI 3	SwCI 4	ScI 5
Task-1 (Test 1)	Y	Y	Y	Y	N
Task-2 (Test 2)	Y	Y	N	N	N
Task-3 (Analysis 1)	Y	N	N	N	N
Task-4 (Analysis 2)	Y	N	N	N	N

Software Control Categories (MIL-STD-882E)

Level	Name	Description
1	Autonomous (AT)	Software functionality that exercises autonomous control authority over potentially safety-significant hardware systems, subsystems, or components without the possibility of predetermined safe detection and intervention by a control entity to preclude the occurrence of a mishap or hazard.
2	Semi-Autonomous (SAT)	a) Software functionality that exercises control authority over potentially safety-significant hardware systems, subsystems, or components requiring a control entity to complete the command function. The system detection and functional reaction includes redundant, independent safety mechanisms to mitigate or control the mishap or hazard. b) Software item that displays safety-significant information requiring immediate operator entity to execute a predetermined action for mitigation or control over a mishap or hazard. Software exception, failure, fault, or delay will allow, or fail to prevent, mishap occurrence.
3	Redundant Fault Tolerant (RFT)	a) Software functionality that issues commands over safety-significant hardware systems, subsystems, or components requiring a control entity to complete the command function. The system detection and functional reaction includes redundant, independent fault tolerant mechanisms for each defined hazardous condition. b) Software that generates information of a safety-critical nature used to make critical decisions. The system includes several redundant, independent fault tolerant mechanisms for each hazardous condition, detection and display.
4	Influential	Software generates information of a safety-related nature used to make decisions by the operator, but does not require operator action to avoid a mishap.
5	No Safety Impact (NSI)	Software functionality that does not possess command or control authority over safety-significant hardware systems, subsystems, or components and does not provide safety-significant information. Software does not provide safety-significant or time sensitive data or information that requires control entity interaction. Software does not transport or resolve communication of safety-significant or time sensitive data.

Chapter 16 – Software Safety

SwCI Matrix (MIL-STD-882E)

SW Control Category	I Catastrophic	II Critical	III Marginal	IV Negligible
1	1	1	3	4
2	1	2	3	4
3	2	3	4	4
4	3	4	4	4
5	5	5	5	5

SCI	Level of Rigor Tasks
1	Program shall perform analysis of requirements, architecture, design, and code; and conduct in-depth safety-specific testing.
2	Program shall perform analysis of requirements, architecture, and design; and conduct in-depth safety-specific testing.
3	Program shall perform analysis of requirements and architecture; and conduct in-depth safety-specific testing.
4	Program shall conduct safety-specific testing.
5	Once assessed by safety engineering as Not Safety, then no safety specific analysis or verification is required.

16 - 41

Level of Rigor (LOR)

RTCA DO-178B

Development Aspect	A	B	C	D	E
Independence Level	High	Medium	Low	Very Low	---
Low Level Req'ts Needed	Yes	Yes	Yes	No	---
Statement Structural Coverage	Yes	Yes	Yes	No	---
Decision/Condition Structural Coverage	Yes	Yes	No	No	---
MCDC Structural Coverage	Yes	No	No	No	---
Configuration Management	Tight	Tight	Medium	Low	---
Source to Binary Correlation	Yes	No	No	No	---
Req'ts Correlate to Target Processor	Yes	Yes	No	No	---
Architecture & Algorithm Verification	Yes	Yes	Yes	No	---
Code Reviews	Yes	Yes	Yes	No	---
SQA Transition Criteria	Yes	Yes	Yes	No	---

← Increasing Work Load & Cost

16 - 42

Chapter 16 – Software Safety

Common Sticky Points

- Establishing the specific LOR tasks

- When, where and how to apply the SCI
  - Is risk really accepted?

- Realizing the SCI can never really change, although the risk probably does

- Hazards and SCI

- Through the successful implementation of a system and software system safety process and LOR tasks, the likelihood of software contributing to a mishap may be reduced.

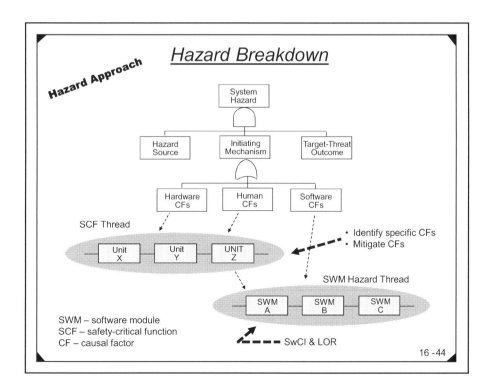

Hazard Breakdown

SWM – software module
SCF – safety-critical function
CF – causal factor

Chapter 16 – Software Safety

Software Fault Tree Analysis

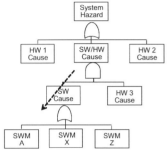

SW hazards are best recognized via FHA and FTA.

FTA shows where the hazard leads into the SW.

DSM = Design Safety Measure

System FTs are useful because they aid in:
- Showing where SW contributes to hazards
- Depicting how a DSM mitigation reduces risk likelihood
    - Should modify FT with an AND gate
    - Mitigate SW hazard with HW

16 - 45

Design for Safety

- Hazards must be mitigated to an acceptable level of risk
- Hazards are mitigated via design / procedures

- LOR tasks improve SW design, but do not directly mitigate hazards
- SW related hazards should have a design mitigation (in HW)
- LOR is insufficient to reduce risk (of hazard)
- LOR reduces risk of SW design

16 - 46

Chapter 16 – Software Safety

Example of SCFs for Tactical Aircraft

- Altitude Indication
- Attitude Indication
- Air Speed Indication
- Engine Control
- Inflight Restart After Flameout
- Engine Monitor and Display
- Bleed Air Leak Detection
- Engine /APU Fire Detection
- Fuel Feed for Main Engines
- Fire Protection/Explosion Suppression
- Flight Control - Level III Flying Qualities
- Flight Control - Air Data
- Flight Control - Pitot Heat
- Flight Control - Fuel System/CG Control
- Flight Control - Cooling
- Flight Control - Electrical
- Flight Control - Hydraulic Power
- Canopy De-fog

- Pilot Oxygen Supply
- Stores and/or Expendables Separation
- Safe Gun and Missile Operation
- Armament/Expendables Ground Operations
- Emergency Canopy Removal
- Emergency Egress
- Ejection Capability
- Landing Gear Extension
- Ground Deceleration
- Structure Capability to Withstand Flight Loads
- Freedom From Flutter
- Stability in Pitch, Roll and Yaw
- Heading Indication
- Fuel Quantity Indication
- Fuel Shut-off to Engine and APU
- Engine Anti-Ice
- Caution and Warning Indications

Source: AOP-52, SwS Guidance for Munition Related Computing Systems (NATO)

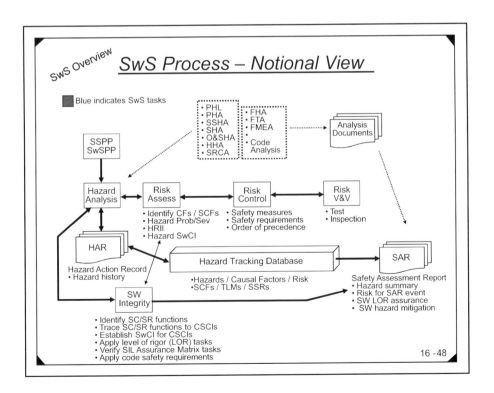

Chapter 16 – Software Safety

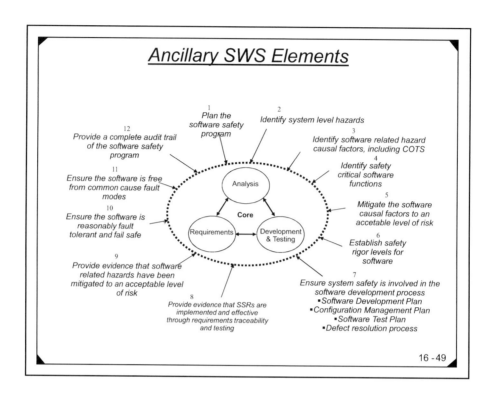

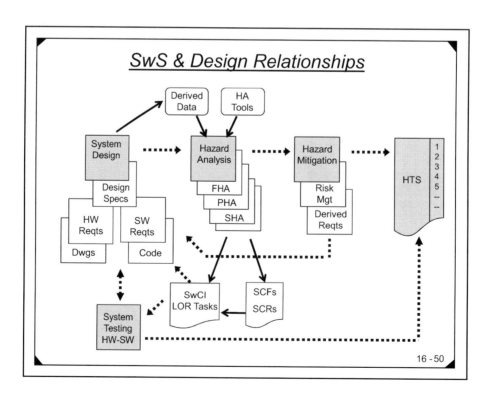

Chapter 16 – Software Safety

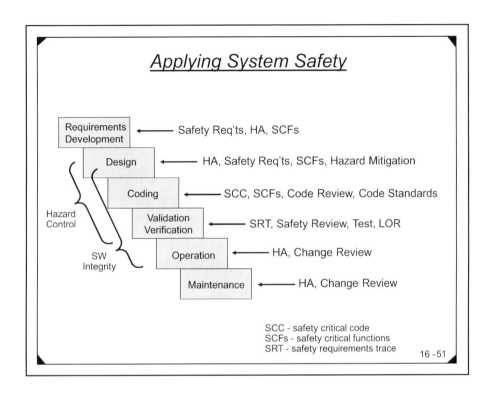

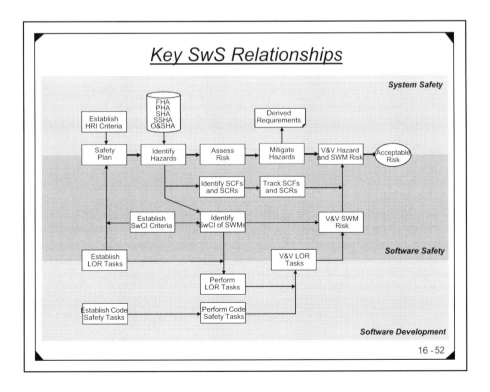

Chapter 16 – Software Safety

Key Software Safety Steps

Step	Description
1	Develop safety plan
2	Collect data (SWMs, Functions, etc.)
3	Identify hazards (FHA, PHA, etc.)
4	Identify / track SCFs (FHA)
5	Identify / track SCRs
6	Identify criticality of SWMs
7	Establish the SwCI for SWMs
8	Mitigate hazards
9	Apply SWM LOR tasks
10	Apply code safety
11	Generate Safety Assurance Case (SAR)
12	Program safety support

SWS Checklist (1)

No.	Element	Why
1	Safety plan	Understand system, criticality, scope, organization
2	Hazards	To know where design is unsafe
3	SRCFs (SW related causal factors)	To know where SSRs are needed
4	Risk Assessment (RA)	To know seriousness of hazards
5	SCPs	To know what parameters are SC in order to protect
6	SCF model	To know what functions and interfaces to protect
7	Code protections	To provide/enhance code protection
8	SSRs	Design requirements to mitigate hazards
9	Traceability – SSRs to hazards	To ensure all hazards are mitigated
10	Traceability – SSRs to design req'ts	To ensure SSRs are present for designers
11	Traceability – SSRs to test req'ts	To ensure SSRs are present and effective (V&V)
12	Traceability – SSRs to code	V&V SSRs and help code maintainers
13	Hazard tracking/closure	To verify hazards are closed and accept risk
14	STR/SPR review/evaluation	To identify and resolve SRCFs
15	Design change review/evaluation	To identify and resolve hazards/SRCFs

Chapter 16 – Software Safety

SWS Checklist (2)

No.	Element	Why
16	Tag SC SSRs	Ensure code maintainers know risk of SW
17	Tag SC software modules	Ensure code maintainers know risk of SW
18	Tag SC software SLOC	Ensure code maintainers know risk of SW
19	TLMs	To relate (understand) risk and CFs
20	SIL for software modules	To identify SC modules and establish rigor
21	Code analysis	Identify hazards and SRCFs in the code
22	Safety precepts	To guide design, hazard identification and SSRs
23	COTS safety	To ensure COTS does not cause hazards
24	Tech Refresh / Tech Insertion safety	To ensure tech changes do not cause hazards
25	Tools validation	To ensure tools do not introduce hazards
26	OS validation	To ensure OS does not introduce hazards
27	Compiler validation	To ensure compiler does not introduce hazards
28	SAR	To document safety case and communicate risk
29	Design support	To coordinate SSRs and safe architecture
30	Test support	To ensure all SSRs are tested and test procedures are safe

Summary

- SwS is a methodology to ensure SW is acceptably safe for use in operational environment
    - A subset of system safety
    - There are many significant differences between HW and SW that necessitates a special approach
- SwS requires two pronged approach
    - Functional hazard control
    - SW development quality control
- Not optimum, but optimistic
    - Assumes SW risk is reduced to acceptable level if certain tasks are successfully performed
    - Based on Software Criticality Index (SwCI)

Chapter 16 – Software Safety

System Safety and Reliability Analysis
Course Notes

Chapter 17
eTree Program

Clifton A. Ericson II
Design Safety Solutions LLC
cliftonericson@verizon.net
540-786-3777

© C. A. Ericson II 2014

Quick Operation Instructions

1. Start New Tree - select *File, New Tree*.
2. Highlight Node - left click mouse on node (text turns red).
3. Add New Node - highlight node, right click mouse, select *Add Node*.
4. Create Gate - add a new node to Basic Event.
5. Save Tree - select *File, Save, Tree (can change file name)*.
6. Quick Save - press floppy disk button at bottom right of tree view screen.
7. Tree Picture Size - press button at lower left *Small/Large* tree view.
8. Edit Node Data - highlight node, click on Node Tab.
9. Edit Tree - highlight node, right click mouse for Edit Menu.
10. Create MOE - give Basic event same name as first MOE event.
11. Create MOB - give Basic event same name as Gate node with MOB branch.
12. Print Larger Text - select *Options, Tree Box Density*.

Chapter 17 – eTree Program

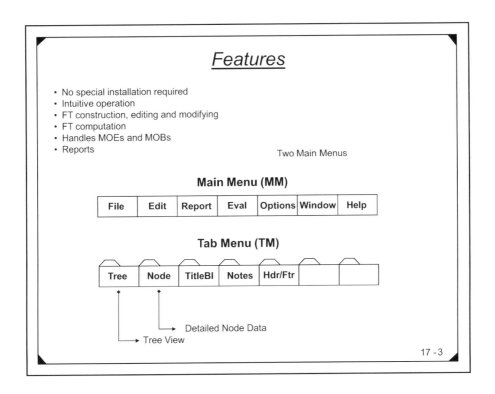

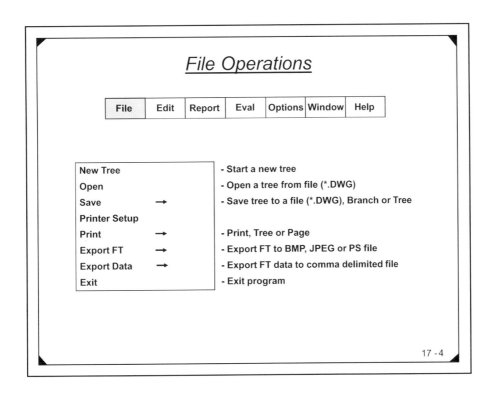

Chapter 17 – eTree Program

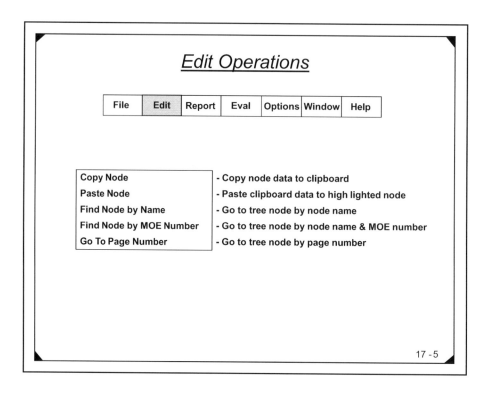

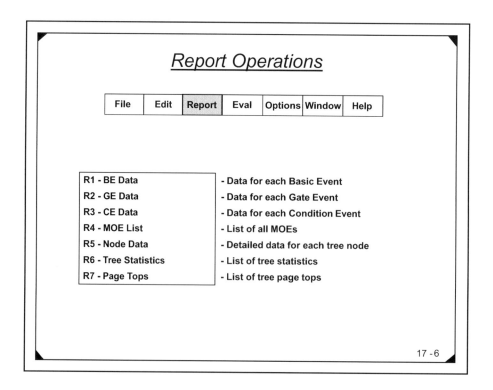

Chapter 17 – eTree Program

Report Operations (cont'd)

| File | Edit | **Report** | Eval | Options | Window | Help |

R8 - Tree Data File	- Displays tree input data file (for latest save)
R9 - Printer Statistics	- Displays printer setup data
R10 - External Transfers	- Lists all Transfer-In trees
R11 - Logic Loops	- Checks tree for logic loops
R12 - Indep Module Check	- Test subtree for MOE independence
R13 - CS/Tree Prob	- Displays Cut Set and Tree probabilities
R14 - Gate/Node Data	- Data for all nodes (different format)
R15 - Unused	-
R16 - Internal Transfers	- Displays list of Internal Transfer nodes
R17 - MOB List	- List of all MOB nodes

Evaluation Operations

| File | Edit | Report | **Eval** | Options | Window | Help |

Generate Cut Sets → Tree	- Compute Cut Sets for entire tree
→ Gate	- Compute Cut Sets for subtree
Calc CS Probabilities	- Calculate probabilities from Cut Sets
Display Cut Sets/Prob	- Display list of Cut Sets/Probabilities
Display System Probability	- Display calculated system probability
Save Cut Sets to File	- unfinished
Open Cut Set File	- unfinished
Build CS Plot File	- Select a CS and build *.DWG file
Calc Gate Probabilities	- Calculate all gate probabilities (*MOE)

Chapter 17 – eTree Program

Node Tab

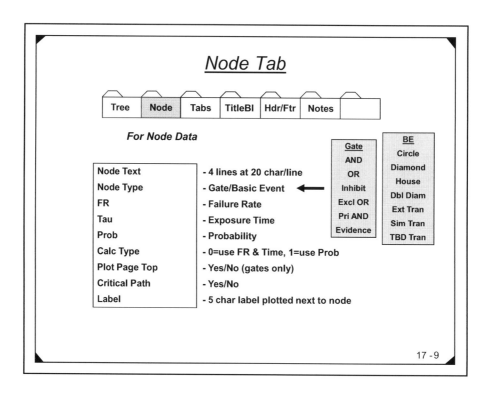

Node Text	- 4 lines at 20 char/line
Node Type	- Gate/Basic Event
FR	- Failure Rate
Tau	- Exposure Time
Prob	- Probability
Calc Type	- 0=use FR & Time, 1=use Prob
Plot Page Top	- Yes/No (gates only)
Critical Path	- Yes/No
Label	- 5 char label plotted next to node

Gate: AND, OR, Inhibit, Excl OR, Pri AND, Evidence

BE: Circle, Diamond, House, Dbl Diam, Ext Tran, Sim Tran, TBD Tran

17 - 9

Three Edit Methods

- Pop-Up Edit Menu
  - Tree Construction

- Node Data Edit Menu
  - Node Tab
  - Node Data

- Node Pop-Up Edit Menu
  - Quick, less data
  - Node Data

17 - 10

Chapter 17 – eTree Program

Pop-Up Edit Menu

Change Name	- Change node name
Reorder Node	- Reorder nodes under a gate
Add Node To Gate	- Add a new event node to gate node
Add New Gate With Node	- Add new gate with input node (into event only)
Add Multiple Event Nodes	- Add N input nodes to a gate
Edit Node	- Go to Node Tab edit screen
Move Node	- Move a node to different location
Delete	- Delete a node, gate or tree branch
Insert Gate/Condition	- Insert a gate above/below or condition
Copy Subtree	- Copy tree branch to clipboard
Insert Subtree	-Insert tree branch from clipboard or file
Repeated Branch (MOB)	- Create or delete a MOB
Page Top Toggle	- Make highlighted gate into Page Top (or undo)

Commands for Tree Construction
(Highlight node, then right mouse click)

Node Pop-Up Edit Menu

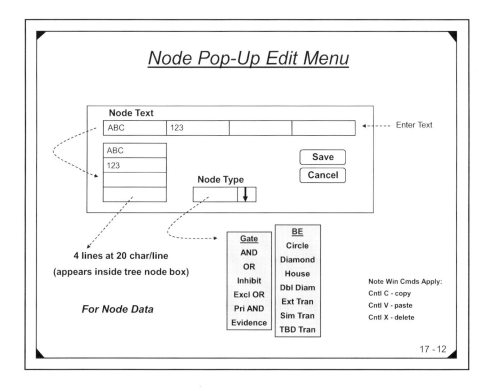

Chapter 17 – eTree Program

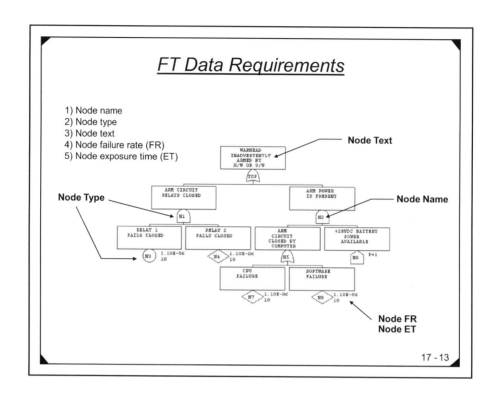

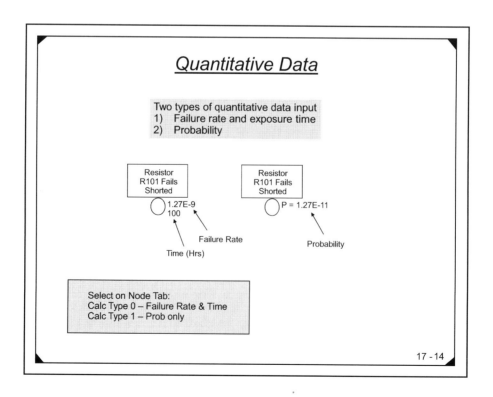

Chapter 17 – eTree Program

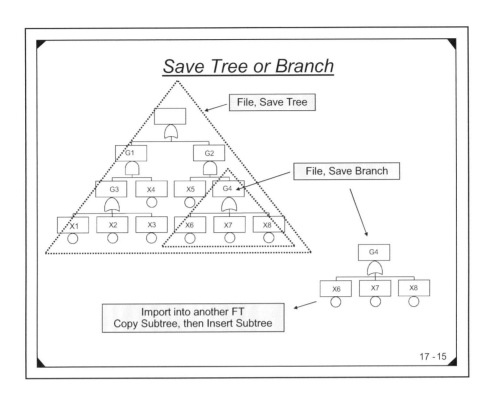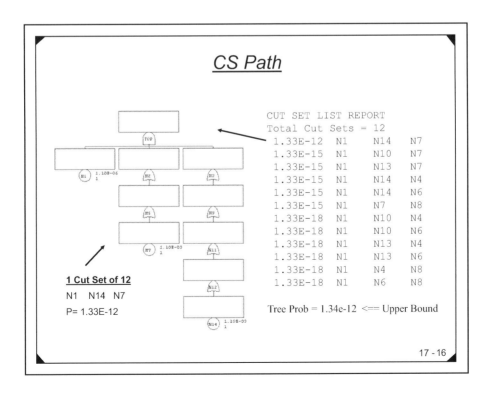

Chapter 17 – eTree Program

Appendix A – References

Safety Books

1) Hazard Analysis Techniques for System Safety, C. A. Ericson, John Wiley & Sons, 2005.
2) Concise Dictionary of System Safety: Terms and Definitions, C. A. Ericson, John Wiley & Sons, 2011.
3) System Safety Primer, C.A. Ericson, CreateSpace, 2011.
4) Fault Tree Analysis Primer, C.A. Ericson, CreateSpace, 2012.
5) Hazard Analysis Primer, C.A. Ericson, CreateSpace, 2012.
6) Software Safety Primer, C.A. Ericson, CreateSpace, 2013.
7) System Safety Engineering and Management, H.E. Roland & B. Moriarty, 1985, 2nd edition 1990, John Wiley.
8) System Safety Engineering and Risk Assessment, N.J. Bahr, 1997, Taylor & Francis Publishing.
9) System Safety: HAZOP and Software HAZOP, F. Redmill, M. Chudleigh and J. Catmur, 1999, John Wiley.
10) System Safety Analysis Handbook, R.A. Stephans and W.W. Talso, 2nd edition, 1997, System Safety Society.
11) Safeware: System Safety and Computers, N.G. Leveson, 1995, Addison-Wesley.
12) Basic Guide to System Safety, Jeffrey Vincoli, Wiley, 2006.
13) System Safety for the 21st Century: The Updated and Revised Edition of System Safety 2000, Richard Stephans, Wiley, 2004.
14) Engineering a Safer World: Systems Thinking Applied to Safety (Engineering Systems), MIT Press, 2012.
15) Assurance Technologies Principles and Practices: A Product, Process, and System Safety Perspective, Dev Raheja and Mike Allocco, Wiley, 2006.

Safety Standards

1) MIL-STD-882E, System Safety, Department Of Defense, Standard Practice, 11 May 2012.
2) ANSI/GEIA-STD-0010-2009, Standard Best Practices for System Safety Program Development and Execution, 2009.
3) SAE ARP4754, Certification Considerations for Highly-Integrated or Complex Systems (1996, SAE Aerospace Recommended Practice).
4) SAE ARP4761, Guidelines and Methods for Conducting the Safety Assessment Process on Civil Airborne Systems and Equipment (1996, SAE Aerospace Recommended Practice).
5) NUREG-0492, Fault Tree Handbook, N. H. Roberts, W. E. Vesely, D. F. Haasl & F. F. Goldberg, 1981.
6) NASA (no number), Probabilistic Risk Assessment Procedures Guide for NASA Managers and Practitioners, August 2002.
7) NASA (no number), Probabilistic Risk Assessment Procedures Guide for NASA Managers and Practitioners, August 2002.
8) DOD Joint Software System Safety Handbook, 2010.
9) DO-178C, Software Considerations in Airborne Systems and Equipment Certification, 2012.

Reliability Books

1) Effective FMEAs, Carl Carlson, Wiley, 2012.
2) The Basics of FMEA, McDerrmot, Mikulak, and Beauregard, Productivity Inc., 1996
3) Practical Reliability Engineering, O'Conner, *3rd edition, Revised*, John Wiley & Sons, Chichester, England, 1996.
4) Handbook of Reliability Engineering and Management, W. Ireson, C. Coombs Jr. and R. Moss, McGraw-Hill, 1996.
5) Design Reliability: Fundamentals and Applications, B. S. Dhillon, CRC Press, 1999.
6) Reliability Theory and Practice, Igor Bazovsky, Dover Publications, 1961.
7) How Reliable Is Your Product? 50Ways to Improve Product Reliability, Mike Sliverman, Super Star Press, 2010.
8) Design for Reliability, Dev Raheja and L. J. Gullo, Wiley, 2012.
9) Human Reliability & Safety Analysis Data Handbook, Gertman, David I. & Blackman, Harold S., John Wiley & Sons Inc., New York, NY, 1984.

Reliability Standards

1) MIL-HDB-217, Reliability Prediction of Electronic Equipment.
2) MIL-HDBK-338B, Electronic Reliability Design Handbook, 1 Oct 1998.
3) MIL-STD-721, Definitions of Terms for Reliability and Maintainability.
4) MIL-STD-756B, Reliability Modeling and Prediction, 18 Nov 1981.
5) MIL-STD-2155, Failure Reporting, Analysis and Corrective Action System (FRACAS), 24 July 1985.
6) MIL-STD-781D, Reliability Testing for Engineering Development, Qualification and Production, 17 Oct 1985.
7) MIL-STD-790, Reliability Assurance Program For Electronic Part Specifications.
8) MIL-STD-975, Standard Parts Derating Guidelines.
9) MIL-STD-1543B, Reliability Program Requirements for Space and Launch Vehicles, 25 Oct 1988.
10) MIL-HDBK-470, Maintainability Program Requirements for Systems and Equipment.
11) MIL-HDBK-2084, General Requirements for Maintainability.
12) MIL-STD-1629A, Procedures for Performing a Failure Mode, Effects and Criticality Analysis, 1980. (http://www.fmea-fmeca.com/milstd1629.pdf).
13) MIL-HDBK-189, Reliability Growth Management.
14) NSWC Standard 98/LE1, Handbook of Reliability Prediction Procedures for Mechanical Equipment, U.S. Naval Surface Warfare Center, September 30, 1998.
15) WASH-1400, Reactor Safety Study, 1975 (mechanical and human failure rates).

Appendix B – Acronyms

ALARP	As Low As Reasonably Practicable
BIT	Built-In-Test
CCF	Common Cause Failure
CCFA	Common Cause Failure Analysis
CDR	Critical Design Review
CM	Configuration Management
CMF	Common Mode Failure
CMFA	Common Mode Failure Analysis
COTS	Commercial-Off-The-Shelf
CSCI	Computer Software Configuration Item
D/I	Death or Injury
DAL	Design/Development Assurance Level
DOORS	Dynamic Object Oriented Requirements System
DSF	Design Safety Feature
ECP	Engineering Change Proposal
EMR	Electromagnetic Radiation
ESOH	Environmental, Safety and Occupational Health
ETA	Event Tree Analysis
FHA	Functional Hazard Analysis
FHRI	Final Hazard Risk Index
FMEA	Failure Modes and Effects Analysis
FMECA	Failure Modes, Effects and Criticality Analysis
FMET	Failure Modes And Effects Testing
FRACAS	Failure Reporting, Analysis and Corrective Action System
FTA	Fault Tree Analysis
HAR	Hazard Action Record
HAZMAT	Hazardous Material
HAZOP	Hazard and Operability (Analysis)
HCA	Hazard Causal Factor
HHA	Health Hazard Assessment
HMM	Hazard Mitigation Method
HMP	Hazard Management Plan
HRA	Human Reliability Analysis
HRI	Hazard Risk Index
HS	Hazard Source
HSI	Human Systems Integration
HTDB	Hazard Tracking Database
HTS	Hazard Tracking System
HWCI	Hardware Configuration Item
IM	Initiating Mechanism
IHRI	Initial Hazard Risk Index
JHA	Job Hazard Analysis
LOR	Level of Rigor
LSSRB	Laser System Safety Review Board
MORT	Management Oversight and Risk Tree
MTBF	Mean Time Between Failure

NDI	Non-Developmental Item
O	Outcome
O&SHA	Operating and Support Hazard Analysis
PDR	Preliminary Design Review
PESHE	Programmatic Environmental, Safety, and Health Evaluation
PHA	Preliminary Hazard Analysis
PHA	Process Hazard Analysis
PHL	Preliminary Hazard List
PLOA	Probability Loss Of Aircraft
PPE	Personnel Protective Equipment
PRA	Probabilistic Risk Assessment
PSSA	Preliminary System Safety Assessment
RAC	Risk Assessment Code
RAM	Risk Acceptance Matrix
SAR	Safety Assessment Report
SC	Safety-Critical
SCA	Sneak Circuit Analysis
SCF	Safety-Critical Function
SCI	Software Criticality Index
SDR	System Design Review
SHA	System Hazard Analysis
SHRI	Software Hazard Risk Index
SIL	Safety Integrity Level
SIS	Safety Instrumented System
SMM	System Mishap Model
SOOP	Safety Order of Precedence
SR	Safety-Related
SRCA	Safety Requirements/Criteria Analysis
SSCM	Software Safety Criticality Matrix
SSE	System Safety Engineer
SSHA	Subsystem Hazard Analysis
SSMP	System Safety Management Plan
SSP	System Safety Program
SSPP	System Safety Program Plan
SSR	Software Specification Review
SSR	System Safety Requirement
SSSTRP	Software System Safety Technical Review Panel
SSWG	System Safety Working Group
SwCI	Software Criticality Index
THA	Threat Hazard Assessment
THERP	Technique for Human Error Rate Prediction
TLH	Top Level Hazard
TLM	Top Level Mishap
TTO	Target-Threat-Outcome
TUE	Top Undesired Event
WSESRB	Weapon Systems Explosives Safety Review Board
ZSA	Zonal Safety Analysis

Appendix C – Exercises

Problem #1

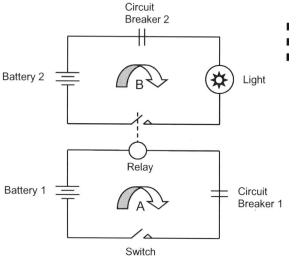

- Construct a FT for this system.
- The Undesired Event is "Light Fails To Operate"
- Ground Rules:
  - Operator closes switch, causing circuit A to close.
  - When circuit A is complete the relay is energized, closing its contacts.
  - The relay contacts are Normally Open (NO).
  - When the relay contacts are close, circuit B is complete and the light illuminated.

Problem #2

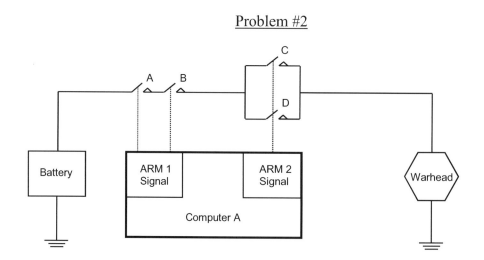

- Construct a FT for this system.
- The Undesired Event is "Inadvertent Warhead Arming"
- Ground Rules:
  - When all of the appropriate switches are closed the warhead becomes armed.
  - The ARM signals are software controlled.

Problem #3

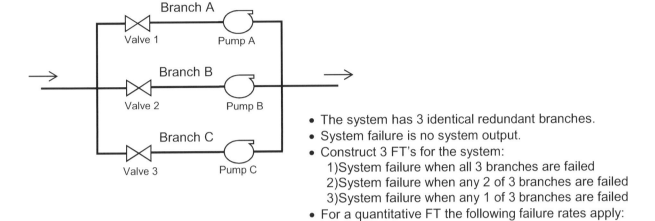

- The system has 3 identical redundant branches.
- System failure is no system output.
- Construct 3 FT's for the system:
  1) System failure when all 3 branches are failed
  2) System failure when any 2 of 3 branches are failed
  3) System failure when any 1 of 3 branches are failed
- For a quantitative FT the following failure rates apply:
  - Valve failure is 1.0×10^{-6}
  - Pump failure is 6.5×10^{-6}
  - Mission Time = 12 hours.

Problem #4

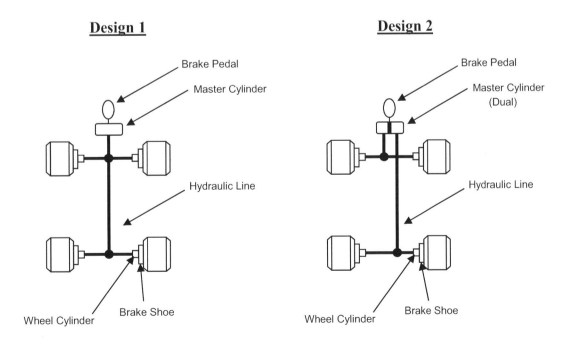

- Construct a FT for these two systems.
- The Undesired Event is "All four brakes fail"

Problem #5

Problem Statement:
- Construct the Fault Tree of the Water Tank system shown below
- The Undesired Event is *No Water Output from P3*
- Derive the Cut Sets and compute the probability of the UE

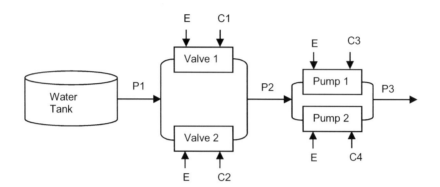

Ground Rules and Assumptions:
1) Each Valve and Pump has its own individual control command (on/off), denoted by C1, C2, C3, C4.
2) Each Valve and Pump has electrical power, which is common to all, denoted by E.
3) Valves 1 and 2 control the supply of water.
4) Pumps 1 and 2 pump the water.
5) P1, P2, P3 represent the three major sections of pipe.

Problem #6

Problem Statement:
- Construct the Fault Tree of the Water Tank system shown below.
- The Undesired Event is *No Water To The Tank*.
- E represents power source; P represents pipes.

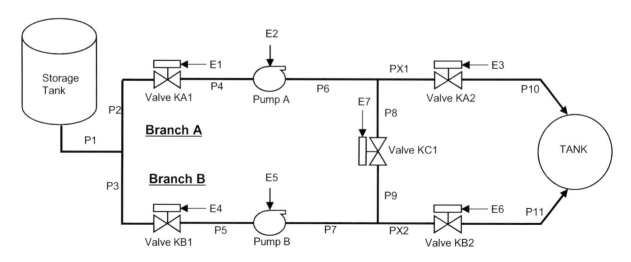

Problem #7

- Derive the Min CS from the following FT using the MOCUS algorithm.

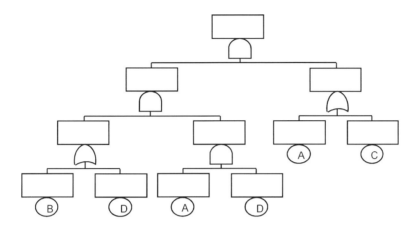

Problem #8

- Derive the Min CS from the following FT using the MOCUS algorithm. Hint: name gates G1 thru G8.

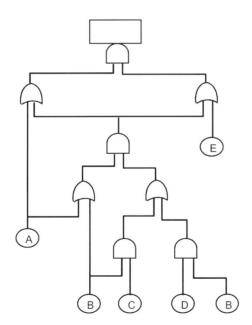

Problem #9

- Prove the following two trees are mathematically equivalent.

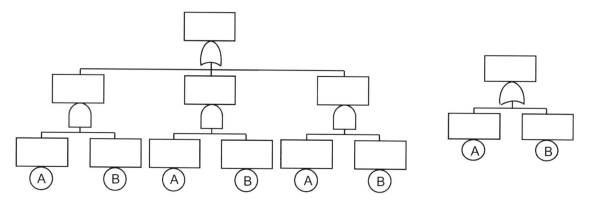

Problem #10

- Construct this FT using the eTree program.

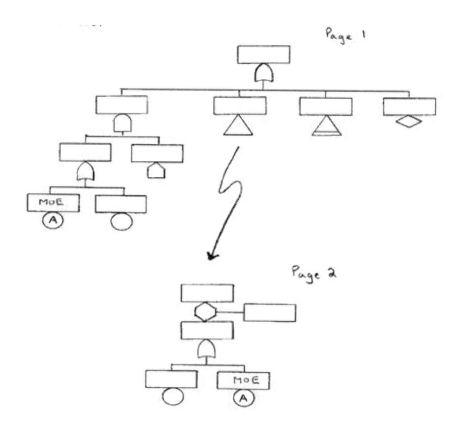

Proof

Made in the USA
Charleston, SC
24 April 2014